TEACHING AND TRAINING

A Handbook for Instructors

Teaching and Training

A Handbook for Instructors

H. R. MILLS

O.B.E., M.Sc. (Lond), C.Eng., M.I.E.E., Dip.Ed.

formerly Director, Science and Engineering Dept., The British Council, and Scientific Adviser to the Commonwealth Relations Office on the Colombo Plan in S. and S.E. Asia; I.L.O. and UNESCO adviser on Vocational Education and Training.

SECOND EDITION

A HALSTED PRESS BOOK

JOHN WILEY & SONS
NEW YORK

Published in the U.S.A. by
HALSTED PRESS
a Division of John Wiley & Sons, Inc., New York

ISBN 0-470-60410-7

Library of Congress Catalog Card No. 72-2045

Printed in Great Britain

Preface

Since the appearance of *Techniques of Technical Training* in 1953, and its first revision in 1967 under the title *Teaching and Training*, there has been a phenomenal increase and great activity in technical training at all levels and in all countries. This has shown itself in the rapid growth in the general literature on the subject and in specialist articles, particularly in periodicals devoted to technical education and training. Training is now a matter of vital national importance in all countries, and is properly regarded as a major concern of educationalists, industrialists, and government departments responsible for economic planning and development. Education and training are recognised as prerequisites of manpower development and economic growth and therefore as a national investment for the future.

The author has drawn on his experience in technical education and training in the developing countries of Asia, where the need for trained manpower is acutely felt and where trained instructors with the necessary skills and knowledge of what to teach and how to teach it are often sadly lacking. This situation calls for special efforts to be made in the field of technical education in these countries, where the many innovations in educational technology are being readily accepted and adapted with greater enthusiasm than in the industrially developed countries. The lives of millions of people depend on urgent economic development and the effective utilisation of the vast human resources of the subsistance economy countries. The developing countries must have more trained teachers and instructors, and in their struggle for economic development they cannot afford to neglect any innovation or opportunity toward the efficient and effective training of manpower.

Owing to the rapid expansion of training programmes all over the world and the demand for more and more trained instructors, there is still an urgent need for a primer for instructors and others who are caught up, without much warning, in the training revolution. Training is concerned, in all countries, with the development of human resources, which is essential for economic growth and the alleviation of hunger and poverty.

H. R. MILLS.

Contents

Introduction – The Importance of the Instructor's Job

This book does not attempt the task of distilling into a small volume the vast amount of work done in educational and instructional technology over the past few years—in the psychology of learning, learning aids, programmed learning, computer-aided instruction, educational cybernetics, curriculum developments, systems approach to training and course design—although passing references are made. It is quite impossible to include in an instructor's handbook a critical account of the many developments in these areas; such developments appear each month or quarter in periodicals devoted to technical education and training, and are dealt with in more advanced textbooks.

It is recommended, however, that training officers and instructors should know what is happening in training methodology by perusing, for example, the brief abstracts of current articles and books on training initiated by the Department of Employment, and selecting for study those that are relevant to their particular job. About 100 short abstracts are published each month by such services including the International Vocational Training Information and Research Centre (CIRF), Geneva, but the instructor who tries to assimilate all the many new concepts in training methods may become quite confused by the jargon used and the apparent lack of agreement among some of the experts.

A number of the innovations may seem unrealistic or remote from situations on the shop floor or training bay. Many of the 3 000 training abstracts published over the past three years are written by psychologists for other psychologists rather than for instructors. To the instructor who makes an honest attempt to understand what it is all about, it may appear that he is being told what he already knows in words he does not understand, or he may feel that the new outlook on training involves so many variables of such complexity that he will prefer to go his own way. The instructor who tries to profit from modern ideas and experimental work may become completely mystified or cynical about some of them.

Instructors should be aware of the scope of present-day research and development in instructional technology, and should, whenever possible, participate in project work concerning training methods and problems relating to their own jobs. Useful experience can thus be gained, particularly through discussions with colleagues and trainees.

Training officers and instructors should keep an open mind for all new ideas and examine them critically; however, unless the results of research can ultimately be translated into simple terms capable of practical application in a real training situation for the benefit of trainees and instructor then time and money is likely to be wasted. Instructional technology must keep in mind its primary objective : to help instructors instruct and trainees to learn.

This book may help inexperienced or untrained instructors to appreciate some of the basic principles of the teaching and training process which still stand, and perhaps prepare them for a fuller participation in what promises to be a sort of educational technology rat race. In this the rats appear to be running in various directions in a three-dimensional black box which enshrines a computerised system of models, with complicated parameters as inputs, and largely incomprehensible jargon as outputs. Practical wisdom is slow to emerge.

The following quotation appears at the beginning of an introduction to an excellent book *'Systems approach to education and training'*, by A. S. Romiszowski :

> 'A system is a little black box
> Of which we can't unlock the locks,
> But find out what its all about
> By what goes in and what comes out.'

Instructors and training officers may be sustained in their efforts to keep in the race by realising that what is commonly known as 'skills analysis' really consists of writing down exactly how a worker thinks, observes and acts in doing a simple skilled operation; the definition of this procedure is given in a glossary of training terms (H.M.S.O.) as :

> 'Skills analysis. The identification and recording of the psycho-physiological characteristics of skilled performance, and the determination of the effector, receptor and decision-making functions involved.'

Skills analysis is by no means new; its value and importance was fully recognised more than 200 years ago.

An account of systematic analytical training in A.D. 1760, relating to the Candlemakers art, was published in the journal *Industrial Training International* (June 1969) and the diagrams, pictures and notes correspond closely to most aspects of modern skills analysis. It is further not without significance that two of the most sophisticated pieces of hardware used in the U.S.A. in educational technology, as computerised instructional aids, or teaching machines, are called after Socrates and Plato respectively, who both over 2 300 years ago, used didactic methods involving systematic programmes of questions and answers and the use of logic akin to those required by modern teaching machines, and programmed learning.

It may help further to reassure instructors, who feel overwhelmed by the spate of literature on instructional technology, if they bear in mind the fact that their job is fundamentally a matter of good two-way communication and sound personal relationships.

The rapid technological developments and industrial growth of the past fifty years have created an urgent and almost insatiable need for efficient and rapid training at all levels, and a corresponding demand for efficient training methods. The old, rigid, time-serving apprenticeships of five or more years, in which young people were expected to pick up their skills with negligible systematic instruction, is now unacceptable.

Many of the new developments in training methods are linked with new media such as films, video tape or computers. Some, such as programmed learning and skills analysis, are, as we have seen, old ideas with a new look, and some are more remote from the instructor's experience and are as yet matters of interest for discussion at symposia of psychologists.

A really significant development is the emphasis now being given to the importance of the learner as an individual, and the special emphasis being given to attitude training and the related behaviourial changes that training can effect. A further development of the past decade is that training can be more effective when emphasis is placed on the techniques of learning in a learning situation rather than on the techniques of teaching in a teaching situation.

There will, however, be no danger that the flood of educational technology will erode the importance of the teacher's work. On the contrary, the teacher, in helping trainees to learn, and creating for them at all times a learning situation, will find that he has a more interesting and arduous task than that of just holding forth to his class. The instructor is now faced with many new methods of teaching and aids to learning, and he is made aware that individuals differ; that one man's learning aid is another man's stumbling block (see p. 213).

Yet the instructor's job is doomed to failure unless there is a desire to learn on the part of the trainees. This desire may have to be promoted and sustained by the instructor, and this may prove to be one of his main tasks, particularly where incentives to learn are weak or lacking. This book emphasises the importance of having the right objectives and priorities in all forms of training. A point which emerges again and again from recently published articles on training is that unless training involves the attitudes and characters of the trainees, it is largely a waste of time.

It is the function of basic technical education to give people a sound understanding of the fundamental principles of science, mechanics and mathematics and an ability to discover for themselves principles and techniques that they will have to master in their work; this can be achieved only through the proper guidance of keen and skilled teachers. Young persons so educated and trained will be able, in a short time, to adapt themselves to any technological environment and learn new skills and techniques. Experiments have shown that there is a good correlation between literacy and ability to learn psychomotor skills. The literate person has learnt how to learn and this ability is transferred to other skills; he is familiar with various modes of communication and has acquired his literacy through the use of several cognitive and motor skills, such as reading and writing.

Training officers and instructors will have a wider range of objectives and greater motivation in their work if they are fully aware of the importance of their task and the valuable part they play in shaping the economy of their country, which is entirely dependent on trained manpower in all sectors and at all levels.

Liaison and co-operation essential

A simple systems approach to training and economic development has been outlined in figure 1 which illustrates manpower production based on co-operation and good communication between education, training (A), industry (B), Government ministeries (D), professional

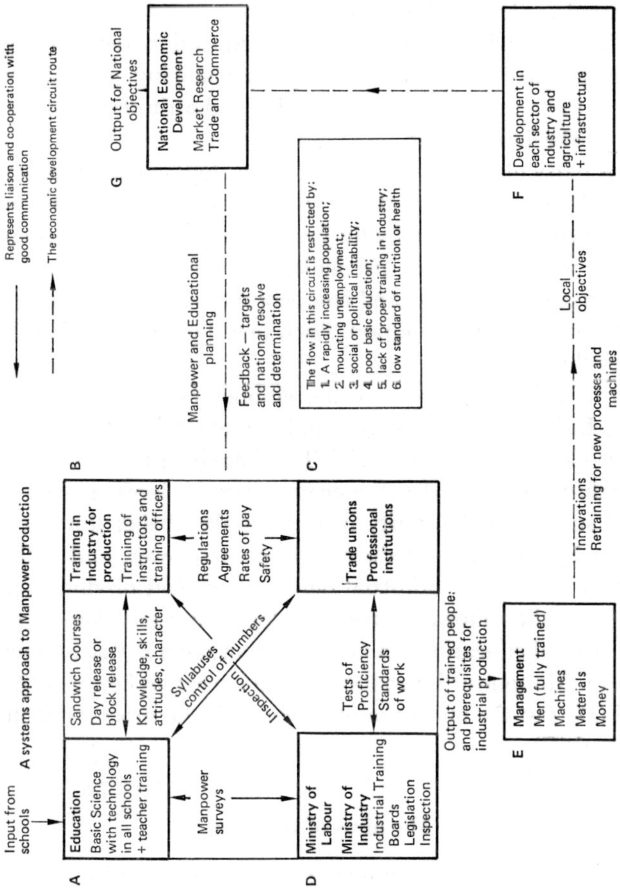
A systems approach to Manpower production
Represents liaison and co-operation with good communication
The economic development circuit route
Input from schools
A
Education
Basic Science with technology in all schools + teacher training
Sandwich Courses
Day release or block release
Knowledge, skills, attitudes, character
B
Training in Industry for production
Training of instructors and training officers
Manpower surveys
Syllabuses control of numbers
Inspection
Regulations
Agreements
Rates of pay
Safety
D
Ministry of Labour
Ministry of Industry
Industrial Training Boards
Legislation
Inspection
Tests of Proficiency
Standards of work
C
Trade unions
Professional institutions
Output of trained people: and prerequisites for industrial production
E
Management
Men (fully trained)
Machines
Materials
Money
Innovations
Retraining for new processes and machines
Local objectives
F
Development in each sector of industry and agriculture + infrastructure
G
National Economic Development
Market Research
Trade and Commerce
Output for National objectives
Manpower and Educational planning
Feedback — targets and national resolve and determination
The flow in this circuit is restricted by:
1. A rapidly increasing population;
2. mounting unemployment;
3. social or political instability;
4. poor basic education;
5. lack of proper training in industry;
6. low standard of nutrition or health

FIGURE I

institutions and trade unions (C). Manpower when produced by this process is of no use unless effectively used by sound management (E) with development plans (F) to produce overall national economic development (G).

To complete the system there must be feedback from (G) to (A), (B), (C) and (D) so that training at all levels is suited to, and controlled by, the needs, either local or national.

The feedback will help to ensure that education and training programmes prepare people for the rapidly changing job market and give them the necessary skills, knowledge and attitudes to profit from further on-the-job training.

This is particularly important in the developing countries where it has been found that only about 50 per cent of the people with a vocational education actually fill jobs for which they were prepared.

What is so often lacking, particularly in the developing countries, is effective communication and co-operation between the groups (A), (B), (C) and (D) in the diagram. Each one should be in close liaison with the other three.

In Britain the importance of this co-operation was recognised by the 1964 Industrial Training Act and the establishment of a Central Training Council and Industrial Training Boards for each of the major industries. In the developing countries (A) is often given undue financial support and priority in the mistaken belief that technical education by itself can produce trained manpower, and will automatically increase the industrial development and productivity of a nation—a condition for industrial development is mistaken as a cause of it. Technical education is a *necessary*, but not a *sufficient*, factor in the system. Manpower is produced by co-operation between (A), (B), (C) and (D), but the output must go through the process denoted by (E), (F) and (G).

The aims of training

The object of this book is simply to suggest ways in which instructors, technical teachers, training officers, works supervisors or managers can improve the quality and speed of training, and to emphasise that their task is not merely to instruct but also to help people to acquire skills, attitudes, habits of thought and qualities of character that will enable them to understand their jobs and perform them efficiently and with satisfaction. This will give them an increasing sense of responsibility in their chosen trades or professions and generally fit them for their part in the world of competitive industry.

Training is not an expensive luxury and reward for the more able or hardworking employees, it is, in this technological age, essential for the proper development of industry at all levels, and is an indispensable factor in the economics of industry. It has long ceased to be regarded as a kind of welfare. It also need no longer be a long-term investment, for with efficient and effective training programmes, the dividends can appear in a matter of months. Training is now essentially a responsibility and a function of management.

An instructors job is to help make full and efficient use of the manpower he has to train, as any wastage in human resources is just as uneconomic as a wastage of time or materials. There is today an increasing demand throughout the world for trained men : scientists, technologists, technicians, and craftsmen. The study of education, technical training and manpower is of such importance to a nation's economic development that it must be regarded as a branch of economics.

Industry quite properly estimates the cost of training—but an equally important complementary exercise is estimating the cost of not training, or the cost of inefficient production resulting from haphazard, prolonged and morale-destroying methods of acquiring skills and 'know-how'. This unplanned type of training is known by the common catch phrase 'sitting by Nellie', a term which admirably denotes the casual and unsystematic learning of a skill by a trainee who merely watches and imitates another worker. In all firms where training is carried out, someone senior and qualified should be given specific responsibility for it. If left to anybody, either it becomes nobody's job, or 'Nellie' ends up doing it.

So important has occupational training become, so encouraging has been the change in attitude towards it in recent years, and so effective the new techniques, that these changes have been heralded as the 'instructional revolution' and its influence is forecast as being comparable in effect on our industrial development, as the Industrial Revolution of 150 years ago.

Those who wish to study the many recent developments in current theories of learning, and in the hardware and software of educational technology, are referred to the select bibliography on p. 254.

Basic training should be broadly based

As developments in science and technology occur they bring to industry new materials, new processes, new principles and new machines; these change the actual jobs that have to be done. In order to meet this very mobile situation, basic training must be broadly based.

In the modern world, the wealth of a country depends on its capacity to produce, and that depends in turn on its possessing a labour force which is highly skilled and very flexible. To raise the level of operator skill and improve flexibility is a pre-requisite of increased production, and for this we need better education and training. Soundly based technical training, in addition to being of immediate importance to local or national industry, can contribute to the raising of living standards in all parts of the world and especially those in the developing countries. It can also increase national wealth more quickly than any other educational expenditure; certainly, in the short-term, more quickly than university education.

There are seven types of training based on the various kinds of trainees, where they are based and whether or not their training is systematic and/or recognised (see figure 2). (3), (5) and (7) are commended; (1) and (4) are generally unsatisfactory.

Wastage through inefficient training

From various recently published reports it is clear that the needs of the large number of part-time students taking advanced courses, and school leavers who are not apprenticed or under any form of systematic training, are being sadly neglected, and a vast reservoir of talent is being wasted. With the rapid technological advances of this century and the need for flexibility, the outmoded time-serving concept of apprenticeship (generally five years irrespective of the nature of the craft skill to be acquired) must eventually be replaced by a more efficient and modern system. The apprenticeship schemes inherited from earlier days were often devoid of regular and systematic training and without any final standardised trade testing. They doubtless served their purpose, supported by the Trade Unions, as a means of controlling the size of the skilled labour force and ensuring security to the craftsmen who had served their time.

A new analytical approach has helped to reduce the drudgery and

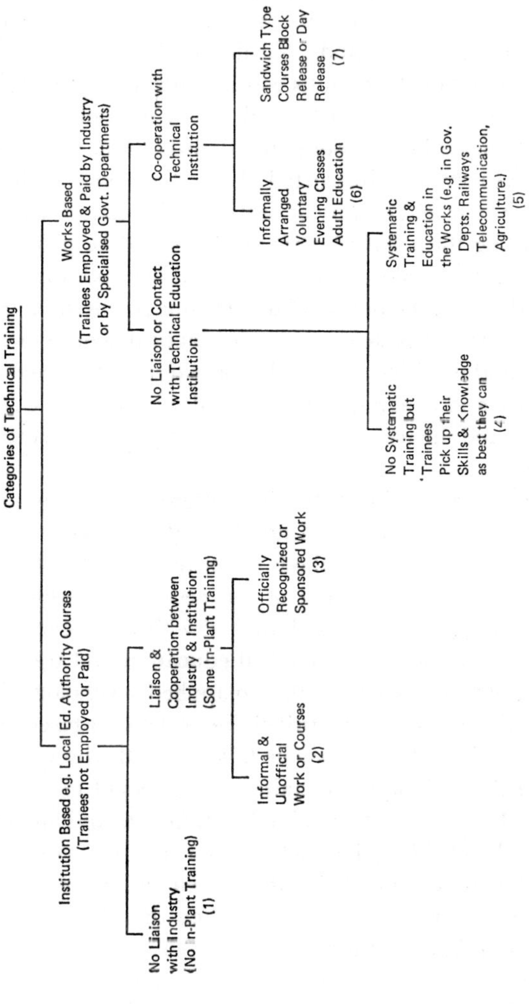
Categories of Technical Training
Institution Based e.g. Local Ed. Authority Courses (Trainees not Employed or Paid)
Works Based (Trainees Employed & Paid by Industry or by Specialised Govt. Departments)
No Liaison with Industry (No In-Plant Training) (1)
Liaison & Cooperation between Industry & Institution (Some In-Plant Training)
Informal & Unofficial Work or Courses (2)
Officially Recognized or Sponsored Work (3)
No Liaison or Contact with Technical Education Institution
Co-operation with Technical Institution
No Systematic Training but Trainees Pick up their Skills & Knowledge as best they can (4)
Systematic Training & Education in the Works (e.g. in Gov. Depts. Railways Telecommunication, Agriculture.) (5)
Informally Arranged Voluntary Evening Classes Adult Education (6)
Sandwich Type Courses Block Release or Day Release (7)

FIGURE 2

inefficiency of old 'time-serving' training, although tradition dies hard. As an example, an industrial psychologist investigated the training of weavers, he was told that two years was the minimum and to be highly skilled, five years. Further, he was told that there was not much hope of success for a weaver unless two of his or her grandparents had also been weavers. The difficulties of the craft were alleged to be many thousands, but the total number mentioned was fifty. A 'museum' of these faults was set up and a systematic training plan put into operation. Girls aged 15–16 so trained achieved top quality in four weeks and were only slightly lower than the very experienced weaver who instructed them; they were well up to the required quantity and quality of output by the end of three months.

The training for a skilled craft or job should be planned and imparted after a job analysis has been carried out and it has been made possible to define 1. the nature of the training required; 2. the syllabus to be followed; 3. the academic and practical requirements; 4. the time, space and machines required for the training in order to achieve a satisfactory and generally agreed standard; and 5. the standard to be achieved, and the nature of the tests to be passed for certification of the trainee.

Education and training authorities often seem to pay more attention to the organisational and administrative problems of training and instruction than to methods of instruction, the techniques of training and the establishment of the necessary learning resources. No attempt is made in this book to deal with the organisational and administrative problems of technical education and training, nor to contribute anything new to studies in applied psychology. It is essentially a handbook of general principles on how to teach, for the instructor who wishes to help people to learn, and who wishes to know more about the techniques of training; it does not tell the instructor what to teach, as he is presumed to be up to date and an expert in his particular subject. Many instructors enter the instructing arena fully confident that they know their subject well and are experts in it, but this confidence can soon be shaken when they have to teach their knowledge and skills to others.

Good teaching or instruction is basically good communication between people, no matter at what level or on what subject. The term 'teacher', 'instructor', and 'training officer' are used here in conjunction with 'class', 'students' and 'trainees', but the essential feature of the relationship in a learning situation is communication, whether this takes place in a works, a college, an office, or on an expedition. Not only is good communication the basis of training, but it is also the essence of good organisation and sound industrial relations. When trouble arises in an organisation or in personal relationships, the cause can often be traced to a lack of communication.

Although the world has undergone many changes, nevertheless, in a letter to the Corinthians St Paul observed : 'Be not deceived : evil communications corrupt good manners,' and so it is today.

Lecturers and new methods

It is well known that an advanced knowledge of a subject in no way ensures the capacity to impart that knowledge. Being taught by a brilliant but boring person has been described as the 'mystification of the ignorant by the indolent'. The most successful teacher is one who is in close touch with the students' own problems and difficulties, and who appreciates that in the early stages of a subject, particularly mathematics, progress by small steps is necessary to achieve understanding, whereas in more advanced stages some of these steps can be 'jumped'.

The word 'obvious' is a favourite with some instructors, but it should be banned or used very sparingly, as what is obvious to the instructor may be far from obvious to the trainee. While it is considered important for school teachers to be trained in the art of teaching, it is still too often assumed that university and technical college lecturers need no training to teach their particular skills. There are not infrequent complaints among students of monotonous delivery, hurried style, no learning aids apart from languid squiggles on a chalkboard, dictated lectures, inaudibility, avoidance of discussion, and most serious of all, the boredom of the lecturer himself. This book is not, however, written with the university or technical college lecturer in mind, but should any read it and be induced to examine their methods and approach, they may care to experiment with some of the ideas and suggestions put forward and improve on them, or at least study the bibliographical references.

More attention to methods needed

A great many books have been written over the past twenty years on the theory and practice of teaching, and on the organisation and administration of technical education and occupational training in industry, but a disproportionately small number has appeared for the guidance of instructors engaged in technical training, many of whom are brought face to face with a class of trainees without any previous experience in the art of communicating skills and knowledge. Although much of the basic material remains, this edition incorporates some of the new developments in the hardware and software of instructional technology of the past few years.

The guidance given is based on practical experience and is intended as a guide to those general principles of training that can be applied to any course of technical instruction, no matter what the subject may be, or for whom the course is intended. As a reference book, it should be useful during a short training course for instructors or technical teachers. Some sections, for example chapters 6 and 7, are designed more particularly for those who have to supervise training, or for those who wish to become acquainted with some of the experiments in practical psychology as applied to technical training.

The instructional revolution from 1950 onwards has come about because of the ever-increasing demand for more and more technical education and training, partly through research and developments in our understanding of how people learn, which leads to programmed lessons, and partly through audio visual aids to learning, which include the use of modern mass media, such as films and television (see p. 213, 'What is involved in teaching and learning'). These developments have proved to be particularly applicable to the teaching and learning of skills.

This book may serve as a stimulus to the instructor who is on the lookout for ideas and new methods. One of the biggest handicaps to a good training scheme is the 'seasoned' instructor who has been teaching in his own way for years. There is a story of one such teacher who applied for promotion on the grounds of twenty-five years' experience. It was rather harshly pointed out to him that his twenty-five years was not fully relevant as it consisted of one year's experience repeated for twenty-four years!

The Services, and many progressive firms, with modern emphasis on technical skills of every kind, accord a high priority and importance to training methods and to the training of instructors. Intense short courses in methods of instruction have been amply rewarded, and suitably planned courses for training instructors of various kinds in industry have enhanced their efficiency and the quality of the training. Twenty years ago the instructor was generally someone designated to look after apprentices and this was often a backwater job for those nearing retirement.

In recent years several institutions for the training of technical teachers have been established in different centres, but these can deal with only a fraction of the number who require training and who would greatly profit from a systematic course on the basic techniques of technical training—of which this book provides an outline. One of the specially valuable results of the Industrial Training Act of 1964 has been the encouragement given to the training of instructors and training officers and the consequent recognition of the responsible nature of their work as a vitally important aspect of management.

Under the powerful stimulus of war, when large numbers of trained men were desperately required, it was found that 80 per cent or more of the skilled jobs in industry could be learned fairly satisfactorily in a fortnight by keen men. By the use of practical and intensive methods of instruction, men and women who had never before been inside a factory could satisfactorily carry out skilled operations after a few weeks' training. Normally such achievements were believed to be attainable only after some years of apprenticeship. Intensive methods are not ideal, but the results achieved helped to focus attention on the value of planned systematic training methods.

A book of this sort cannot be expected to turn all bad instructors into good ones, but it is nevertheless essential at the start to refute that mischievous doctrine that a good instructor has some sort of divine gift denied to other men. The 'gift of the gab' does appear to be inborn in many people, but this must not be mistaken for ability to instruct. Instructors are in fact *made* not *born*. There are techniques of instruction, which have to be studied, learnt and practised. The more common errors have to be brought to the notice of all who aspire to proficiency. For example, instructors on the whole tend to talk too much. They frequently indulge in over-long periods of continuous talking as this practice requires less preparation and teaching skill than planned class activity. It requires less effort and skill on the part of a teacher to foster a teaching situation than a learning situation. The instructor's job ultimately is not simply to transmit information or to demonstrate, but to help his trainees to learn. Principles of class comfort are often lost sight of, and visual aids, including films, are often used ineffectively because they are used without discrimination or skill. It has been shown by carefully conducted experiments on actual classes under instruction that the speed and success of technical training depend very largely on the methods of instruction. There is a good deal more in a piece of good instruction than meets the eye.

Teaching is in fact hard skilled work for which study, training and practice of all the tricks of the trade are necessary for the full efficiency which will give satisfaction and a sense of achievement both to the teacher and the taught. Good instruction makes difficult subjects easy to learn, but bad instruction makes even easy subjects difficult to learn.

Thousands of courses are being organised all over the country in technical institutions, firms, and government departments and specialists are frequently called upon to run or help with such courses at short notice, having had little or no training for the work, and often with little previous experience as instructors.

Although the techniques used in training require a good deal of pedagogical wisdom and experience, the whole subject is really a common-sense application of those principles of good communication and man-management which have proved to be most effective in achieving the specific objectives of training, and particularly the larger objectives of all training, namely to make the man on the course a better and happier man at his job, to improve teamwork, and to increase production.*

* It is interesting to note that the term 'production training' is used instead of 'practical training' in the syllabuses of technical colleges and schools in the U.S.S.R.

I

The Background to Training and Some Objectives

Introducing the new worker

Training in industry begins from the moment the trainee arrives at the gates of the works. He must from the start be made to feel that he is joining a team, and that he has an important part to play in that team. To ensure this the company should be seen to be interested in the applicant and should obtain in advance, if possible, all the relevant facts about a candidate before any interview takes place. This is conveniently done by means of an employment application form, which not only helps the interviewer, but serves as a basis of the man's record and future progress reports (see chapter 6). Forms vary considerably depending on the grading of the post, but usually include :

1. Personal particulars—name, address, sex, age, marital status, children, physical fitness, etc.;
2. Education and training;
3. Type of job required, alternatives, when free to start;
4. Previous employment, rates of pay, references;
5. Hobbies or special interests.

This form should be clearly worded and kept as simple as possible. If completed at the works, someone from the personnel department should be available to give guidance and answer queries.

Early impressions are important, and so the reception should be courteous, and any directions given should be clear. Attention to detail will be amply rewarded : for example, the waiting-room for receiving prospective trainees should be reasonably well furnished, and equipped with a mirror and writing materials for filling in any forms necessary. There are several good films on how to treat people, including new workers.

A new man selected for training should at the earliest opportunity be given an idea of the whole purpose of the works or industry, its relation to everyday life and its importance to the community.

The induction course

A new employee should if possible be given an induction course during the first few days, to build up a feeling of confidence in the firm and its management. The course should make clear how the trainee's work will fit into the whole organisation (good visual aids, charts and diagrams will be of great value), and would include the following:

> Welcome by company official. A short history of the firm. How it is organized. How departments are connected. Who does what at the top. The firm's products and how it serves the community (for example in an automobile works, a trainee should be given a chance to ride in the final product and to appreciate accurate and skilled workmanship). The works library. Connections with other firms, and with other countries. The regulations, including those on health and safety, why they are made, and why they must be kept. Canteen, welfare, and recreational facilities. Social and sports clubs. Advancement and promotion, and how to set about further special training or education, for example, under the local education authority. Possibilities of transfer to other jobs in the firm or outside. Whom to go to if in difficulty. Outline of the training course to follow. Explanation as to how wages are calculated, what statutory deductions are made, and whom to approach in case of error. (It is surprising how often firms fail to explain such matters and later pay a high price for this failure.)

Some instructors, recollecting their own experiences on joining, feel it incumbent on themselves to 'cut all new arrivals down to size' and to put them in their place. This is unwise. Each trainee probably joins the firm with some elementary skill and knowledge relevant to his future job, which he has learned at school, from his father or through evening classes. This may not be of any great importance to the firm or the training officer but the instructor should show reasonable interest in any achievement and should be encouraging. Encouragement to learn and motivation are fundamental prerequisites to learning. Let trainees come to the dirty, unsavoury and tough parts of their work in the natural order of events. It is bad psychology and therefore bad training to give men hard meaningless jobs when trying to teach them a technical subject.

The new trainee should have an early opportunity of meeting those at the top responsible for his selection and ultimate advancement and also some of those with whom he will eventually be working. Ensure that he gets to know the right group, and does not fall into bad influences at the start. The whole object of training can easily be undermined by the born grousers and the disintegrating elements in the shop or factory who regard newcomers as a heaven-sent audience. A word or two in the right quarters can generally put into operation a sponsor system, whereby a man of influence, happily adjusted in his own life and in the job, can be encouraged to look after a beginner. In larger firms the training officer will naturally keep an eye on young workers and gain their confidence so that they feel they can go to him in any difficulty.

Confidence in the top or the administration is essential for trainees no less than for workers. The training of employees is ultimately a responsibility of management and demands the most careful study and attention. Training in its fullest sense includes further education, widely based, so that the employees become more skilled in their work, happier on the job, and are made aware of opportunities for advancement or even for changing their training should they want to.

A great many semi-skilled jobs are ignored in training schemes in industry as not being worth the trouble. This attitude is not justified by experience. Planned training for all levels of skill is of great value in industry, not only for the improvement shown, time saved, and reduced accidents, but because a man's morale is enhanced by his being given a course. He is happier and therefore works better if he feels the firm attaches some importance to him and his job.

The quite common practice of putting a learner to work by the side of, and under the vigilant eye of, an old hand is not always the best way. The old hand may be quite unsuited as an instructor. He may not be inclined or be able to give the necessary time to the task, or he may have undesirable tricks or short-cuts that are better elminated than perpetuated. He may even be reluctant to pass on the tricks of the trade. He may enjoy his superior skill and may unconsciously wish to maintain his superiority unchallenged.

Training for the semi-skilled has been described as 'Kate learnt from Jane, and Jane learnt from Betty and Betty learnt from Nelly and Nelly never knew'—or happened to be left-handed! The subject of the training of supervisors has increasingly engaged the attention of management in recent years, and in the future it is essential that proper attention should be devoted to the systematic training of semi-skilled and unskilled operatives.

Many of the problems in industry can be traced to lack of proper training:

1. delays from errors and mistakes (bad communication);
2. faulty work returned;
3. excessive wear and tear on equipment and machines;
4. untidiness of work area;
5. specifications not followed;
6. quality standards not met;
7. carelessness over personal safety;
8. materials not stacked properly;
9. ignorance of safety rules;
10. lack of interest in work;
11. importance of minor jobs not stressed;
12. discouragement because work cannot be mastered;
13. lack of substitutes for absentees.

The larger objective of training in industry

Sound training imparts not only a way of doing, but also a way of thinking, so that a trained man confronted with a situation acts wisely and quickly. For example, he will deal with a complaint or a problem effectively. 'Work out all the facts.' 'What is the right course?' 'Is it honest? Is it free from selfish ends?' 'Is it tactful?' 'Is it going to make for better understanding and teamwork?' He will plan with a clear head. 'What do I want to achieve?' 'How can I achieve it?' 'What are the snags?' 'How can I overcome them?' 'Time?' 'Materials required?'

A man's training goes on throughout his employment with new jobs to do, new machines to handle and new people to manage. The instructor's job is not merely to instruct, it is also to train men, and training involves much more than the acquisition of skills and facts; it involves habits of thought, relations with the boss and with fellow workers, reliability on the job, initiative, leadership, in fact everything that goes to make up a skilled, competent, reliable man, with an influence for good in the shop or office. This aspect of training has taken on a special significance in recent years and has a variety of new names such as sensitivity training, group training, human relations training. (See Chapter 4, p. 67.)

All this may seem too ambitious or idealistic for the ordinary instructor, but it is essential that this larger objective of training be kept in mind if he is to carry out his full responsibilities effectively. This approach should make the instructor's work a vocation rather than a routine job.

Some essential qualities are caught and not taught

Many problems which are accepted as being technical are in reality moral problems. For example, in a certain naval training establishment during the Second World War, the care and maintenance of batteries in landing craft were taught efficiently, resulting in a high percentage of the class passing out with good marks. Operational reports, however, showed many failures, which were attributable to neglect by men who, technically, had high marks on the course, but who were not up to the mark on reliability, sense of duty, or responsibility. Care of batteries, like thousands of other jobs in industry, is not only a technical problem, it is also a *moral* one. Foremen and supervisors or works managers are all well aware that a man can be trained so that he knows what to do, how to do it, and why it has to be done, but even then the man may fail to do it—through lack of a sense of responsibility, laziness, or lack of consideration for other people.

Training should turn out men committed to a right course of action. Training is concerned with character and ethical issues. It is of little use to preach or to teach team-work, reliability, sense of duty, and so forth; these qualities are *caught* and not *taught*—caught from the environment, and particularly from the instructor. This aspect of training is most important and must be appreciated before the techniques of training can be properly assessed. To teach responsibility you must give responsibility.

John Wellens in *The Training Revolution* says, 'If we think the national problem of occupational training is soluble by the mere multiplication of formal training courses, we shall not succeed. Formal courses correspond to the head, but also involved is the heart. Beside the know-how, there is the attitude. What a challenge to any person who has responsibility for the work of others!'

Many young workers in industry have received the minimum of state education but they can, through progressive training and encouragement in their jobs, gain competence and confidence which will help to develop their personalities and wider interests more fully perhaps than a longer period at school could have done.

This aspect of training, with attention to the larger objective, has been much neglected; it was totally obscured by the Industrial Revolution, with the factory system and division of labour, but closer attention to it will help in the solution of the social problems of our time. A story from pre-Industrial Revolution days is apposite here: three stonemasons at work were independently asked, 'What are you doing?' The three answers received were: 'Cutting this piece of stone'; 'Earning three shillings a day'; 'Helping Sir Christopher Wren to build a cathedral'.

A man is not truly happy unless he finds an outlet for his personality and energy through his work. In the actual day-to-day training programme, we shall see that opportunities will present themselves for the instructor to develop in the class a sense of loyalty, teamwork, initiative, a sense of duty, cheerfulness and perseverance in the face of difficulties.

It is these qualities which bring men to the top of their professions, and opportunities should be provided to develop them during training. The instructor is called on to be a leader and an example. Smart stories of how you used to cover up a bad piece of work or pass the buck on to the next process, may illustrate your technical point, but they do not contribute to the larger objective of training.

Training within industry

In recent years it has been found that good supervision increases productive efficiency, and supervision can be improved with training.

The Ministry of Labour sponsored for some years a scheme known as *Training Within Industry for Supervisors* or T.W.I., which succeeded in demonstrating the inadequacy of the old, casual, time-wasting, unsystematic way of learning a job and showed how the speed and quality of training could be improved on by adopting systematic methods and programmes, and training instructors in the techniques and skills of teaching and learning. T.W.I. continues but its functions have been largely absorbed by the work of the various Training Boards under the Industrial Training Act. Nevertheless, in industry a heavy share of the burden of technical training still falls on the foreman or supervisor who must be capable of 1. instructing new workers (job instruction); 2. handling men and dealing with problems of human relations (job relations); and 3. maintaining a scientific attitude to methods of work and the problem of efficiency (job methods).

The training is largely concerned with giving supervisors a right outlook and attitude towards their important responsibilities; the method of group discussion (p. 161) is extensively used. Instructors should be given every opportunity to keep pace with new technical and organisational developments, and to this end refresher courses should be run wherever necessary.

Workers selected for part-time instructional duties should not suffer any financial hardship through loss of bonus or overtime as this would be a poor reward for their being selected as instructors.

You teach not only a subject but a person

To keep the aim of training in view, it helps to remember that you are teaching a subject to a person with a mind, a will, a personality, and a character. In the past decade a great deal of research has been carried out into the social, economic and technical problems arising from automation. A subject of study has grown up called ergonomics which management can ill afford to ignore in its practical applications, as it deals with the effects of machines on man. It is a subject of growing importance in these days of dependence on machines and automation.

How we learn

An instructor must have the *attention* of his class. The only way a person's mind becomes aware of, or grasps, a fact or a situation, is by *attending* to it. We attend to situations we are *interested* in. A tiger suddenly appearing at the window would instantly demand our attention, because for some special reason (some call it the instinct of self-preservation) we are intensely interested in such a situation. Our self-preservation is clearly involved, and we shall see that whenever a basic instinct is linked with a situation, our interest is guaranteed, and attention is inevitable.

Some instructors endeavour to secure attention by shouting to keep the class roused, or by threatening the class with harsh words or an eagle eye. 'Attention' so achieved is of little value. The trainee puts on an air of attending as a sort of defence mechanism, with a glassy stare fixed on the instructor and a mind miles away.

Some instructors gain the attention of their class (attention of a sort) by making jokes in season and out of season. Beware of being the funny man. Jokes and amusing anecdotes about the work have their place, and are often a good means of rousing interest, as we shall see later, but the great danger is that although attention is secured by the class waiting for the next joke, interest tends to shift from the subject of instruction to the instructor-comedian.

Attention through interest

Attention is the responsibility of the instructor, or to put it more tersely, 'inattention is the instructor's sin'. The instructor's job is far more than standing before a group and delivering a series of correct statements about the subject to be taught. Lectures, talks, discussions, demonstrations and practical exercises must be made as interesting as possible by all the devices in the instructor's stock in trade, as will be shown in chapter 3.

Our interests derive their power from our fundamental instincts or their emotions. Consider, for example, how popular films and novels appeal to a major instinct such as sex or hero-worship. Contrast the ease and pleasure with which a boy, for example, can rattle off a list of Cup winners with his painfully poor performance in, say, listing the Kings of England—although the history master has done his best for weeks to induce the boy to assimilate this knowledge. Hero-worship is strong in boys and amounts almost to an instinct, so that they are vitally interested in their hero's batting average or the fortunes of the Arsenal's captain. They identify themselves with their heroes. Their self-esteem is enhanced (the self-regarding sentiment is, or should be, strong in us all), they are interested, therefore they attend; the impression goes deep and they remember the facts of football.

Memory is good only where there is attention through interest. In fact it is a good adage for the instructor to bear in mind that there is no such thing as bad memory, there is only bad attention. In technical training, a strong motivating force which makes the trainee want to learn, and which supplies a background for interest, is this self-regarding sentiment which we shall see in chapter 4 takes a variety of forms and shows itself as pride in one's work, the pride of the craftsman, the love of creating and the instinct of self-display.

We like to regard ourselves as good useful people, and feel bad about failure and loss of face. All these feelings are generally heightened if a man has a family, and perhaps parental responsibilities.

Pride in one's work and desire to 'get on' are right and proper motivation or 'motivators' and ensure a measure of interest in a dull subject, even with a boring instructor, for whenever there is interest, no matter how obtained, there is attention. A man who does not have a well-developed self-regarding sentiment and has no urge to get on or to express himself in his work is a real problem, not only to the instructor, but also to himself.

Many of the subjects taught under technical instruction are initially not sufficiently interesting in themselves to hold the attention of say a random group of young men, unless the subject matter can be linked with a major interest which acts as a motivator. What spurs a trainee on, especially to do private study or voluntary practical work, is the thought he commendably entertains of himself as a qualified mechanic, or a chartered engineer, and the loss of face entailed in failure. This self-motivation can, and often does, operate in spite of an unimaginative or even boring instructor. An enthusiastic and skilled instructor can, however, make his subject *intrinsically interesting*, so that the trainee develops an interest in the subject over and above the interest which comes from his self-esteem and the desire to get on. The drive from the desire to get on and the pleasure from intellectual exercise and achievement reinforce one another. (Incentives and motivation are dealt with more fully in chapter 4.)

Basis of instruction principles

The instructor must understand the larger objective of industrial technical training, and keep it in mind at all times.

1. The class must be motivated.
2. They must be encouraged to want to learn.
3. Our innate desires to get on, to be thought well of, to belong to a community or to help our family, are mainsprings of our conduct.
4. They motivate and ensure interest.
5. Interest commands attention.

The various steps in a piece of instruction

We may now anticipate the next few chapters which deal with the basic principles of instruction in the order in which they have to be dealt with by the instructor. These steps can be recognised in any piece of instruction, whether it be a lesson, lecture or demonstration.

1. *Object.* Know what you are aiming at. The object of training must be kept in mind. The object of a particular piece of instruction must be clear and definite, and appreciated by the trainees.

2. *Preparation.* The effectiveness of the instruction depends on the preparation. The key question is 'How does this help the learner to learn?'

3. *Presentation*—putting it over. The instructor may be regarded as the producer, stage manager, star, scene-shifter and prompter.

4. *Reception*—taking it in. 3 and 4 are not really separable in a good learning situation. They can be replaced by the single term 'communication'.

5. *Assimilation.* The class should take an active part in instruction and inwardly digest what has been presented, by using it.

Technical training must be a part of a man's education. Apprentices soon become set in their ways. Always be on the lookout for opportunities to develop character. A man can be highly skilled in his job, but if he is unco-operative or indifferent to the fate of his colleagues in the firm, unreliable and unpunctual, he is a possible danger in the works—in spite of his technical ability.

6. *Assessment.* This is concerned with testing, conducting examinations, and progress reporting. It discloses weaknesses in instruction and can be said to increase motivation.

7. *Feedback.* Students should know their successes and mistakes without delay. Experience gained throughout a lesson should be fed back into the next lesson plan.

2

The Instructor's Job

The success of a piece of instruction and the rate at which learning takes place depend on the preparation. The more you plan and prepare, the more you help the trainee to learn. This fact is taken advantage of by the development and growing use of modern methods of instruction, such as programmed learning, which depend for their success on the planning and preparation of specific programmes, with careful thought being given to all the principles discussed later in this chapter.

A Ministry of Education statement of October 1962, and valid for all time, for teachers and instructors, states: 'No teacher has ever wasted the time he has spent on the vigorous analysis of both subject matter and operations of thought, or the sources and types of errors in learning—all of which are forced on him by the programming operation; indeed in some sense this kind of programming lies at the very heart of all successful teaching and should be the concern of all teachers and those who train them.'

We shall see in a later chapter (p. 152) which deals with the planning and preparation of a course of training, that the modern approach to training begins with a systematic analysis of the factors involved: the needs of the organisation, the problems to be solved, the cost, and the training objectives in terms of what a trainee will be expected to know and to be able to do, and how he will be expected to respond to the working situation and problems on the job. These objectives are often called behavioural objectives.

The people for whom the training is intended (trainee target) and what they already know on entry (entry behaviour) must also be analysed and taken into account. The analysis should be a cooperative effort of management, training officers, supervisors and instructors.

Preparation for the instructor includes preparation of the *subject-matter* to be taught, its selection and arrangement; preparation of *demonstration material* apparatus, diagrams, etc.; and preparation of the *minds of the class* so that the new work follows on from previous work, facilitating understanding and a logical growth of knowledge of the subject. Good preparation will clarify what you are going to say, what you are going to demonstrate, and what the trainees have to do before, during and after the lesson. Finally, it is important that you should plan what tests you are going to apply to help you to find out how successful or otherwise your efforts and the trainees' efforts have been.

Your work should be so well planned that your students get the impression that you mean business and know your job, and that you expect them to do their best work and co-operate. Give them at the outset an overall picture of the course—why it is important and how you propose to run it, with a clear statement of the main and subsidiary objectives.

This chapter considers briefly some of the basic principles of teaching as these must be observed at the preparation stage, but these principles are dealt with more fully in subsequent chapters. Have a clear idea of what you are going to teach and wish to achieve.

Here are examples of instructors at their work. (This could be put on either as a playlet or recorded on tape, in a course for instructors, to illustrate the difference between the instructor who is prepared and has a clear objective and an instructor who has no clear objective. Incidentally, in any course for instructors, it is an excellent idea to introduce such variations into the programme to show how much more effective they are than mere talk and chalk. See chapter 8, p. 167). A tape recording is easy to make and play to the class. It can be rehearsed with a prepared script with local touches and some ad lib dialogue. The theme in this example is that of a works manager or head of a technical department who, in making a tour of inspection, comes across a class of apprentices just about to commence instruction. He asks the instructor what the course is and what particular piece of instruction is about to begin.

The replies may be as follows.

a. I am going to give the class something on machine tools.

b. I am going to carry on with batteries.

c. We are going to continue the electricity course for motor mechanics.

or

i. I am going to demonstrate how the angle of the cutting tool affects the quality of the turning, and each member is then going to try out various angles for themselves.

ii. This morning we are going to break down some concentrated acid and measure acid density and discover how this changes with temperature.

iii. We are now going to trace the circuit in the cut-out and voltage regulator and practice making adjustments on an actual car that is not charging properly.

a, b and c had only vague and generalised objectives; i, ii and iii had clearly defined and limited objectives.

A class, from either enthusiasm or boredom, will often try to lead you away from your objective even when you have it clearly defined and in mind. If you do not know exactly what you are trying to do there is little hope of your doing it, and you can never test what you have done.

The objective

The instructor's first task in preparation is to get quite clear in his own mind what the objective of the training is in general, and what the objective of each period of instruction is to be.

The general objective may have several aspects which are often closely interrelated.

1. To *motivate* the trainee to want to learn. To stimulate interest in his particular job, and in his future.

2. To impart the necessary *knowledge* of his work including background, theory, and related subjects.

3. To impart the practical *skills* and techniques of the work.

4. To train men in *attitudes* to work, principles of team-work and right human relations so that the proper conditions for willing and loyal co-operation are encouraged between all personnel in all departments.

6. To provide the means for making a clear *assessment* of the trainee's progress in and aptitude for a particular job so that a useful grading of the men results, a proper selection of men for advancement can be made, and both the instructor and trainee know how they are getting on, i.e. plan for the feedback. Reference to the model on p. 29 might be useful at this point.

All these aspects must be kept in mind at the preparation stage, but careful consideration must also be given to the specific objective of each piece of instruction or lesson. Your main objective here should pass the test of being *practical*, *useful*, *definite* and *limited*.

The specific objectives in a piece of instruction can be of several kinds; for example, understanding a principle; understanding how a thing works; problem solving; how to study and work on one's own; correct way of doing a thing; and achievement of speed and accuracy in a 'motor skill'.

What to teach

This may be defined by an analysis of the needs of the organisation or by a set examination syllabus. However, as the instructor should know much more than he is expected to teach, it is a common mistake to attempt to impart too much information, particularly too much detail, in the limited time allowed. As a result trainees often carry away a confused idea of dozens of facts—possibly important to an expert (who can, and often does, check them in reference books), but not essential in a first course. It is far better to teach a little thoroughly than attempt a lot superficially. Basic principles are more important than details, and teaching for understanding is better than teaching mere facts. You should, therefore, work out the actual content of your instruction, that is, the things to be included, by applying a system of priorities on a simple plan.

In all technical subjects there is an essential core of knowledge or skill that a trainee *must* know. Make this the *must* of your instruction. Then there are the valuable and interesting parts which a trainee *should* know if there is time. Finally, there are the trimmings of the subject that are often elevated for no apparent reason to important places in the syllabus. Details concerning the weight of parts of a machine, for example, are often included, whereas the man who has to lift them has a much better idea of the weight than can be conveyed by a statement that the part weighs 20 lb or 9·10 kg.

The frills of the subject are often a great temptation to an instructor who feels he must display a superior and detailed knowledge at all times. The trainee *could* know these details if time permitted, without detracting from the proper stress being laid on the *must* and *should* parts. Training programmes are invariably limited by time, and therefore the total amount of matter that might be taught should be sorted out into these three categories : what the trainee *must* know, *should* know, and *could* know (figure 3).

Seemingly unimportant points in the *could* know category can sometimes be raised in priority on the grounds that they rouse interest, or motivate the class or contribute to the larger objective of training.

In practice, the instructor should make a list of points in order of priority with the time to be devoted to each.
When the time has been used up draw a line. You will have to leave out items low in the scale of importance.

FIGURE 3 Priorities in a training programme

A simple example

Suppose you have ten minutes in which to teach a friend how to handle a motorcycle because of some emergency. He can ride a cycle and drive a car, but has no previous experience of a motorcycle; your list might look something like this:

	Time allotted in minutes	
Sitting on, and getting feel of machine	½	*Must know* 8
Switching on petrol and starting engine, throttle and air controls, including demonstration by you	1	
Clutch and gear changing (machine on stand)	1	
Brake operation, and safety precautions	½	
Actual practice on road under supervision	5	
Lights, horn	½	*Should know* 2
Checking oil and petrol. Position of tool-kit, cleaning plug	1½	
Place of: puncture repair outfit, pump		*Could know*
Cleaning carburettor jet, magneto contacts, fitting spare bulbs		

Sequence in presentation

Although the presentation of a lesson is dealt with more fully in Chapter 3, it is necessary to consider here the sequence in which material is presented, as this sequence must be thought out in advance, and is therefore a part of preparation. Always plan and work with a *learning situation* in mind. For example, consider what questions you will ask, and what questions you anticipate from the class. What questions can you encourage them to ask? What problems can they tackle?

A few minutes should be spent in introducing your lesson, making its purpose clear and arousing the interest of the class. Sometimes a story, a personal experience, or a press cutting can be used to advantage, but one important rule is : *start within the trainee's experience.* For instance, in an elementary course on electricity, start with examples of domestic appliances, before going into complicated appliances used in the works. Adjust the tappets on Smith's motorcycle before dealing with the tuning of a large diesel engine.

Prepare the class for new material, which should be 'grafted' on to the old. Revise what should be known, so that you are sure of the foundation on which you are building. Revision should be clear and methodical. Most subjects have a logical development, and any gaps in previous work seriously hinder progress. This is particularly true in mathematical subjects in which progress depends on a good grasp of fundamental principles.

Now suppose that, with the object clearly in mind, you have decided *what* to teach. The next step is to arrange the matter in the order in which it should be taught. The sequence rules for a learning situation are well known to schoolmasters, and also apply to technical instruction. They are :

proceed from the known to the unknown,
from the simple to the complex,
from the concrete to the abstract,
from the particular to the general,
from observations to reasoning,
from the whole to the parts, and back to the whole.

A course on the gyroscopic compass, for example, could begin with a toy top or gyroscope with which the class is familiar. It is more instructive to observe, collect evidence, and get the class to deduce a principle or rule, than to state a rule and then set about finding examples.

A sequence which serves instructors well in teaching a great many simple skilled operations is: 1. prepare and analyse; 2. explain; 3. demonstrate; 4. class imitate; 5. practise (with proper emphasis on 4 and 5, as we shall appreciate later); and 6. test.

As a general rule, an intricate mechanism should be understood as a whole in very general terms before being analysed into its various main parts; it should then be *built up* bit by bit into its main parts, and finally shown fully assembled.

Sequence is mainly a matter of common sense, but it is not invariably true that a strictly logical sequence is the best one. For example, a logical way to teach a language would be to begin with the alphabet, followed by words, grammar, and sentences, but a trained language teacher does not follow this method today. Language students usually start by attempting to speak meaningful sentences. A child learns to speak a language before learning the alphabet and the grammar. Thus, the psychological order for teaching some subjects may have precedence over the rational order.

Stages in learning a skill

The stages in learning a skill follow the normal systems approach or scientific method.

1. Define the objectives in terms of what is required to be done and what must be known in behavioural terms.

2. Analyse the skill in detail and record the operations and the sequences. It is important to understand the pattern of the movement as a whole and then practise it in parts by systematic instruction and demonstration.

3. Demonstrate the skill with a running commentary, paying special attention to the key points and the more difficult parts.

4. Let the trainee try for himself as soon as possible, practising the skill either in parts or as a whole, initially under guidance.

5. Assess the effort; praise or correct behaviour and explain, if necessary, how things went wrong. Test progress and let the trainee know the result. A sign of increased skill in many jobs is whether a trainee is able to do it without looking; also increased speed is an essential factor in skill acquisition for production.

Some skills are best acquired through what is known as a 'whole' method, by which a complete task or operation is practised from beginning to end. Others are best dealt with by a 'progressive part' method, in which the learner practises one part at a time until he can achieve an acceptable standard of speed. Any two successive parts can then be done together to the required speed, after which a third is added and so on until the whole job can be done in one operation. A third method is to practise each part in isolation to the required standard and then combine them.

The relative efficacy of these methods has been found to vary with the type of skill. Experiments in training men to operate a turret lathe showed that those trained by the isolation and progressive part methods made better progress than those who learned by the whole method. Others conducted on training men in a variety of skills show that there is no particular value in learning disjointed material (such as the various complicated operations to be tackled in working a turret lathe) by a whole method, but there is real advantage in learning closely linked operations in this way. The effectiveness of the part method in teaching complicated skills seems to result from its allowing variation of the time spent on each part according to its degree of complication.

In technical instruction, a point frequently overlooked by instructors is that the mind needs frequent pauses and rests even in a forty-minute period of instruction. This will be dealt with more fully in chapters 5 and 7, but as far as sequence is concerned, it is worth remembering that planning a lesson is rather like writing a letter. The main point of the letter should be made either at the beginning, after a suitable opening, or at the end; in the middle, it is liable to be regarded as relatively less important.

It is interesting to draw an analogy between the learning process and the process of digestion. You cannot eat non-stop for an hour, but you may manage satisfactorily a banquet of five or six courses, with the usual intervals between courses. Your instruction should not be without pause of some sort. You must allow time for the class to digest mentally what they have taken in, and present the 'courses' in the psychological order, in much the same way as a chef plans a menu. The analogy stops here: a good dinner usually leads up to the main dish, the *pièce de résistance*, somewhere in the middle, but in an instruction period, it is generally found that interest, and therefore attention, is at a low ebb in the middle of a period. A graphical representation of how the attention of a class being subjected to a talk or lecture fluctuates with time is shown in figure 4. Interest starts off high, though no matter how good an instructor may be, he will rarely have the full attention of all the class. Interest drops. Fatigue and boredom set in and persist, until banished by the pleasant anticipation of the end of the period. The real attack on the main objective of the lesson should therefore take place when attention is at its highest; that is soon after the start, or else at the end. It is here that you arrange the *must know* part of the instruction. Of course there are exceptions, and the shape of the curve in the hands of a good instructor can be modified almost at will. Your pauses 'between courses' for digestion can be made by making blackboard summaries or presenting the difficult points in a different way. Questions provide an opportunity for class participation and assimilation. These pauses should occur every ten or fifteen minutes.

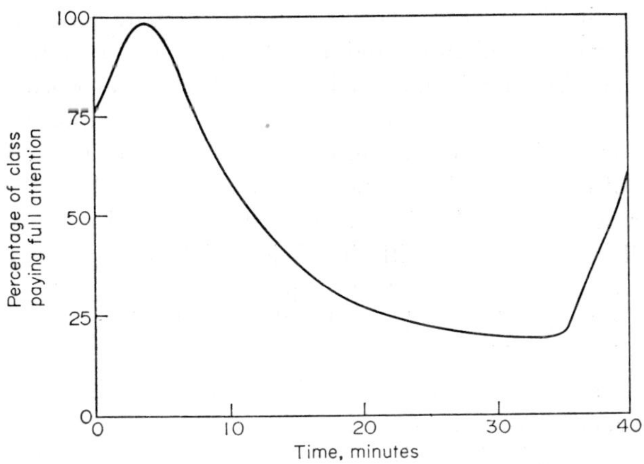

FIGURE 4 Class attention: fluctuation during lesson

Prepare the end of your instruction so that you leave time for questions from the class and time to hammer home a final summary, definite and clear, bearing directly on the object of the lesson. Leave time if possible for the class to give back what they have taken in. 'There is no impression without expression': you do not know a thing until you are able to explain it to someone else. There is no need to elaborate this point; it is relentlessly borne home to instructors whenever they attempt to teach a piece of work for the first time. If you have a trainee who is sure he knows all about a topic (but you are less sure) suggest to him, as a test, that he tries to explain it say to an uncle who knows nothing of the subject.

The previous knowledge of the class

Proper preparation for a new class is possible only if you know what stage in the subject the class has already reached. It is most unwise to assume that a man knows nothing. One often hears the old-time instructor remark 'I always assume they know nothing. I know where I am then.' This may be sound in teaching men a special works process that the trainees could not possibly have learnt elsewhere, but it should not be assumed that previous instruction given by others is of poor quality and must therefore be done again. It is easy for an instructor to catch out a class on a few small points, and then deduce that they know nothing about the subject. Such an attitude is disloyal to other instructors, and there is nothing more boring to a class than going over familiar ground. If a class is a little rusty on a subject, do not dismiss the whole of their past instruction as worthless. Recapitulate patiently. Learning and recalling will be rapid the second time and will give the class some confidence in themselves and in you.

It is of course equally wrong to assume previous knowledge that you are not sure the class has acquired. In the practice of instruction we shall frequently meet with the wisdom of the middle way between two extreme views (see chapter 3, p. 48).

The best course is to find out what the new class knows. This can be done either by asking questions at the first opportunity or, alternatively, as soon as the class is assembled, by a test, either written or practical, depending on the subject of instruction. The test can be, for example, a simple alternative-answer type test, which can be marked in a few minutes and the results declared. A somewhat lighthearted test given by the author to a newly assembled class of instructors is reproduced on p. 225. The class took the test on the first day of the course and, without any warning, at the end; in all cases there was a marked improvement, which incidentally is a demonstration of how to produce a feeling of achievement. Try it now, and then repeat the test after reading the rest of this book.

What pace can the class stand?

A valuable rule has already been mentioned on p. 14, it is better to teach a little well than a lot badly. A man can take in only a limited amount in a particular time; increasing the pace or the lecturing time, or the number of periods, will not necessarily increase the amount learnt and assimilated. There may even be a decrease in the overall achievement if a mental saturation point is reached.

Another good rule is that it is better to go at the pace of the bottom half-dozen in the class rather than the top half-dozen. The reason for this is that if you go at the pace of the brightest, those at the bottom will be lost, and will go from bad to worse; if you bring the slower group along with you, the top will certainly be with you. You should save them from boredom by giving them more advanced work from time to time or getting them to assist with the practical demonstrations. In a large class you may find it best to go at the pace of the middle 50 per cent and give some extra attention to the bottom 25 per cent and additional, more advanced, work to the top 25 per cent. Whatever the pace, the instructor should always be fully prepared in his subject. The class sees through a shallow teacher who is 'one page ahead' of the class. It is a good thing for an instructor to be in close touch with his students' difficulties, but not *that* close!

In the preparation of material every opportunity should be taken of improving your own knowledge of the subject. It is very easy in these days of rapid technical development to be out of date, or unaware of some recent advancement that the class has heard about through books and periodicals or television. You must be able to stand up to all reasonable questions on and around the subject. You will not go far with your class unless you have their respect and confidence, as a man who 'knows his stuff'.

'Sense appeal'

You now have the objective of the whole course and the objective of the lesson clear. You have found out just what the class knows, you have a good idea of what the class can take in at one time and at what pace you can go. You have arranged the subject matter in order of importance and decided what the class *must* know. At this point you are ready to deal with the actual problem of devising ways and means of 'putting it across'.

The best way *may* be a straightforward lecture, consisting of forty minutes' talking from you, but this method is more likely to meet with success in a course for an arts degree at a university than in a technical course in a factory or technical college, where the training must be practical. Men have to be taught to work and act, and therefore the best way of instructing them is to use methods that actually train them in the work they have to do. Methods of training that are remote from the practical application of their skills and knowledge are largely a waste of time for the class and the instructor. In chapter 8 various kinds of instruction are considered in detail.

We have five well-recognised senses: *sight, hearing, smell, taste* and *touch,* plus a kinaesthetic sense which is a sense we are all aware of although it is often explained in words we do not understand. These senses are the gateways or channels to the mind, and all play a part in the learning process and in the techniques of instruction. The table on p. 111 shows that, in the training function, teaching, learning and the assessment of knowledge, skill and attitudes have to be applied in the right 'mix' for optimum effect for any particular objective.

The right mix is generally a matter of common sense and experience but it is worth thinking about at the preparation stage. The rule is, use as many of the senses in your instruction as you reasonably can. Smelling and tasting do not come into many branches of technology, but whenever the materials of a subject have a distinctive and harmless smell or taste, let the class smell and taste them. Cookery, chemical engineering, food technology, safety courses and sanitation, might well use smelling and tasting in their courses. Some important information on these subjects cannot be conveyed in any other way.

The sense of *touch* plays a very important part in many parts of technical instruction. We often have occasion to talk about the *feel* of a thing, and whenever it is relevant the trainees should be allowed and encouraged to handle things that have a characteristic 'feel'.

The qualities of surfaces, and abrasives, can only be explained by feeling. Suppose a lady wanted to teach you the difference between a pair of nylon and a pair of rayon stockings. A verbal discourse on the subject would convey practically nothing; samples of each could be produced, and the difference between nylon and rayon could be *shown* to you. You would still have no clear idea of the difference.

The real difference is to be appreciated only by feeling a pair of each, and as soon as you have experienced this, the sense of touch, sight and hearing join together in making the distinction clear to you.

The kinaesthetic sense provides information and experience about the actions and reactions of our limbs, their positions and the forces we exert in using them. This sense plays a dominant part in all practical skills as it is concerned with the precise positioning of hands, fingers, feet, limbs, trunk, and the reactions in the connecting muscles, and neural paths to the brain. Although difficult to describe and explain in words we can readily call to mind many examples from our own experience of the kinaesthetic sense. One example is the handling of the controls of a car. We do not have to look where the controls are, and we know the 'feel' of each of them in operation. We know the right pressures to use and the sequence without conscious effort. A golfer will spend hours getting his hands, arms, stance just right for a successful drive or putt.

The chart on p. 25 shows how we learn through our senses, and incidentally provides a list of aids to training or audio-visual aids generally available (see chapter 10).

Theory and practice: knowledge, skill, attitudes

We are all familiar with the division of instruction into theoretical and practical—often unfortunately spoken of as if they were constantly in opposition. It is often said in a derogatory way that such and such a person is not much good as he is a 'theoretical' man as opposed to a "practical' man. This is an unfortunate distinction as there is no real opposition between theory and practice in any technical subject. If there *appears* to be, then the theory is not correct, or it has not been properly applied. If the full and proper theory be applied in detail to any practical problem, then perfect agreement will be reached. *Sound theory is sound practice conscious of itself, and sound practice is sound theory unconscious of itself.* The three types of instruction can be recognised by their aims (see figure 18, p. 111).

1. Imparts facts and *knowledge*, in an organised manner, about a technical subject. This is the usual theoretical side of instruction.

2. Imparts *skills*, which involve muscular dexterity, and co-ordination of mind and muscle. This is essentially the practical side of instruction, but knowledge and skills cannot be sharply distinguished,

as the ability to perform a skill and the ability to handle facts and form concepts often merge into each other. This is not surprising since the brain controls every aspect of knowledge and skill. A combination of knowledge and skill is sometimes referred to as a technique.

3. Imparts *attitudes*, ways of thinking and general behaviour, particularly in group situations. Training that does not produce men committed to a right behaviour pattern in attitudes is of little value, and can be dangerous.

During training we have to help trainees acquire a blend of knowledge, skill, and attitudes that will give them and the organisation a means of achieving their objectives—with satisfaction (see p. 111). An example will illustrate this more clearly.

A man decides to take up driving a car. He buys a copy of *How to Drive a Car*, and a copy of the *Highway Code*. He learns all about the internal combustion engine, starting, stopping, throttle, clutch, brakes. He learns the rules of the road and all the signs. He acquires considerable knowledge.

Next he is given a practical demonstration in a car by a friend, and is allowed to 'practise' in a large field. He starts the car, changes gear, turns round and acquires considerable skill and kinaesthetic sensitivity in handling the car at various speeds, still in the large field. All would agree that he is not yet ready to be let loose on the public and allowed to drive the car in London, say from Marble Arch to Fleet Street in the rush-hour.

Skill in handling the mechanical parts of the car, a knowledge of the car and highway code, and an ability to be courteous to fellow road users, have to be fused together in actual experience into a way of behaving correctly in various circumstances. This produces a good driver; that is, one who has acquired knowledge, skill and a right attitude. We shall now see how *knowledge, skills* and *techniques*, each have their own needs in regard to 'sense appeal' in instruction.

Knowledge, involving facts, theoretical principles, should be taught using as much appeal through the eye as possible. It has been estimated that we gain about three times as much technical information through our eyes as we do through our ears, and hence the emphasis must be on good models, instructional diagrams and good blackboard work.

A boxer knows that one in the eye is worth three in the ear. This holds good in instruction. At the preparation stage always try to in-

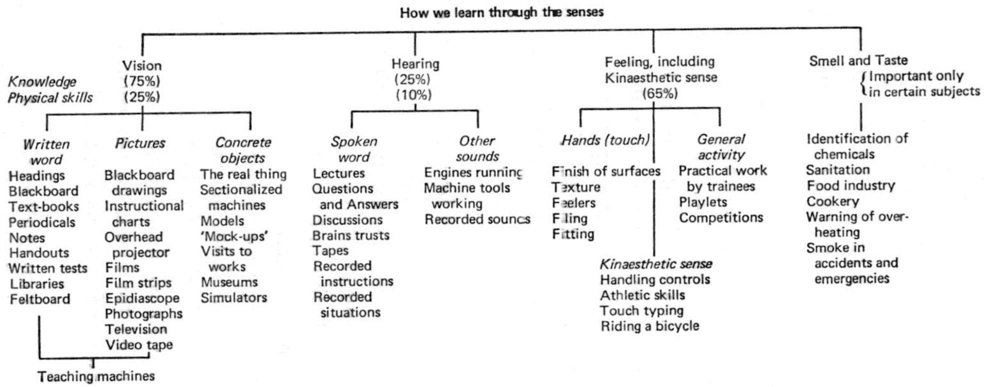

Use this diagram to see if you are making the maximum appeal to the senses when preparing for instruction. Keep the class as active as possible. In skills we learn by *doing* – but doing with progress and satisfaction. Select your method and means of testing to suit the objectives and the class, (see 'What is involved in teaching and learning').

FIGURE 5 How we learn through the senses—a list of points to training

crease the appeal to the eye, and minimise the actual talking. I believe it was Confucius who said : 'one picture is worth a thousand words.' Use posters, the blackboard, charts, film-strips, models, or the real thing. Ears nevertheless have a valuable part to play (p. 156).

However it should be borne in mind that valuable as good visual-aids are, they tend to encourage passivity. You should, whenever possible, use activity methods or, as they are sometimes called, *participatory* methods. The words 'I see' are often used to be synonymous with 'I understand'. There is a warning here for instructors : because you have shown or demonstrated something, it is no guarantee that it has been understood.

In teaching a *skill* from scratch, the emphasis must be made on *learning by doing*—using muscles, co-ordinating mind and muscle, developing the 'thinking hand' as it has been called. As we have seen, the kinaesthetic sense plays a dominant part in learning a skill. In each case the instructor"s voice plays only a small part, but nevertheless an important part. This is not an attempt to belittle the instructor; he is, at all times, of supreme importance in helping trainees to learn. You can only learn a *skill* by *doing* it. The following operations are *skills*, and can only be learnt by *doing* : armature winding, turning, riding a bicycle or a horse, skating.

In teaching a *skill*, talking should be severely restricted to a few points such as why the skill is important (to motivate the trainees), safety precautions, and any essential mechanics that will help in the learning. After that, *talk and chalk* must give way to *learning and doing* by the trainees, with the instructor more or less gagged but standing by to help where necessary.

During the war, an investigation was made of the way technical instructors, teaching physical skills, actually spent their time, and how they *should* spend their time for best training results. It was found that instructors on the whole explained matters for 59 per cent of the time, demonstrated for about 25 per cent of the time, and allowed the class to practice for 16 per cent of the time, as shown in the diagram (figure 6).

This is probably a fairly common allocation of time used by a great many instructors. By means of scientifically planned experiments with groups of men, using some control groups to eliminate any disparity in the learning rates of different groups, it was found that a class learnt a skill best when explanation, demonstration, and practice were mixed in the proportions : explanation 10 per cent, demonstration 25 per cent and practice 65 per cent as shown in figure 7. This confirms the point already made that an instructor generally includes too much talking in his programme.

In learning a *skill*, it has been fairly well established that practice, or the actual use of hands and muscles, is nearly three times as important in the time-table as demonstration or using the eyes; and that explanation, or using the ears, is of comparatively little importance in the learning process.

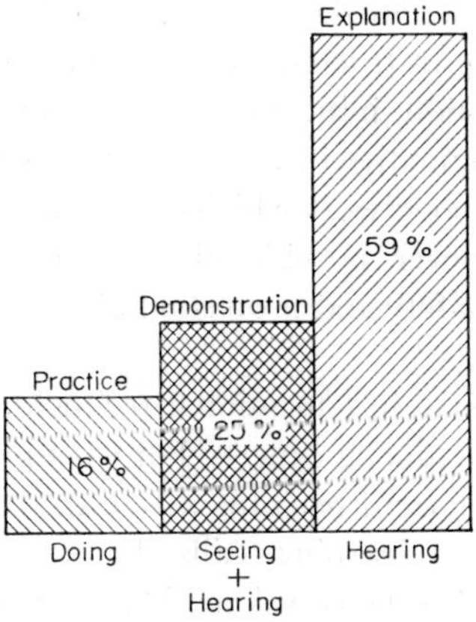

FIGURE 6 Teaching a skill: bad use of allotted time

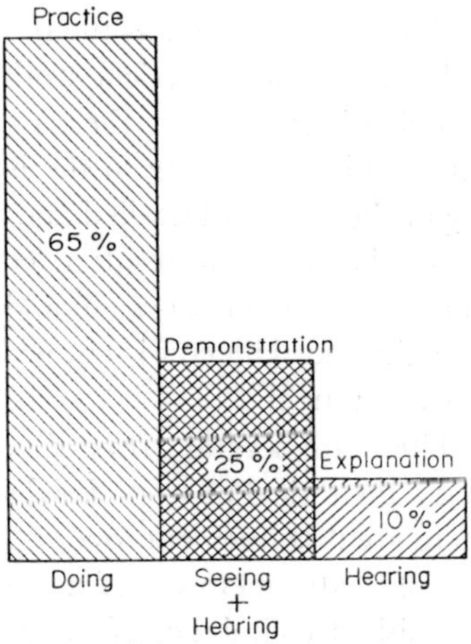

FIGURE 7 Teaching a skill: good use of allotted time

Sixty years ago typewriting was taught by the sight method, it being assumed that the habit of looking at the keys would cease as speed increased. This proved to be wrong for the habit became ingrained and could be overcome only with great difficulty, thus hindering the acquisition of automatic skill. Once the training objective was clarified as typing with the eyes free to follow notes or shorthand, instruction was planned from the start with this goal in view and a screen was placed over the keyboard. This ensured learning by touch. Within a few months students could type at three times the speed achieved by the former method, and with greater accuracy and less fatigue. The kinaesthetic sense had been effectively used.

This example illustrates the need in training of a specific goal and a purposeful and direct method of achieving it. Methods of instruction must be adapted to the objective; flexibility in approach is one of the marks of an efficient and trained teacher.

In learning a complex skill (for example, driving a crane), which, as we have seen, generally depends on knowledge, physical skills and habits of thought, the important thing is *practice*. Explanation and demonstration can take the trainee a little way along the training path towards efficiency, and no further. Beyond that, explanation and demonstration become a waste of time for all concerned. Skills are acquired only through experience, so that when preparing a piece of instruction on a skill make plenty of provision for the men to 'have a go' as soon as the basic knowledge has been acquired.

In laboratory practical work, instruction sheets often show hard conscientious work on the part of the instructor, but are so drawn up as to leave nothing to the imagination, initiative, or understanding of the trainee. Detailed and unimaginative instructions on how to proceed at all points result in 'cookery book' work which can be rather boring. Practical work should whenever possible give scope for thought and skill, consistent with the safety and protection of expensive equipment.

There are several areas of controversy in educational technology and a great deal of scope for discussion among experts. For example, there are disagreements over the efficiency of teaching machines, the use of texts, methods of teaching and learning skills. Even well established techniques of questioning can be overdone. Does a programmed system of questions fired at the class help or inhibit learning by leading trainees by the nose? It is the mark of a progressive and keen instructor to question training practices and principles in order to discover the limits within which these are valid.

An instructor has to consider at the preparation stage of his instruction five broad sets of factors :

1. What are the objectives?
2. For whom is the instruction intended?
3. What teaching methods suit 1 and 2?
4. What learning aids will best suit 1, 2 and 3?
5. What means of testing will be objective, valid, reliable and comprehensive?

He generally has very little choice over 1 and 2, as they are decided by the organisation, but he is free to select from 3, 4 and 5. A rough analysis of each of these factors is given on p. 213 and the possibilities listed under each heading show the wide scope of instructional technology and what is involved in the instructor's job.

Checking a Few Points at the Preparation Stage

Matter, Materials, Minds

Keep the overall *objective* of training in mind. What opportunities are there for attitude training?

Is the immediate objective of the instruction clear, practical and limited in behavioural terms?

Does it fit in with the syllabus?

Check question	*Decided by*	*Notes*
what does the class know already?	questions, or a rapid test	this should be settled well beforehand if possible
what shall I teach?	**MUST** *know*	job breakdown and analysis, note key points
	Should know	arrange matter in an order of importance first and then divide
what method?	could know	Is it practical; useful?
how fast can I go?	continuous assessment	go the pace of the slower members, but keep the bright ones busy, and plan participation for all
is the *subject-matter* prepared?	instruction notes	—
	headings and summaries	what are you going to say?
	precis	what questions are you going to ask?
are the *minds* ready to receive it?	previous work well assimilated	revise if necessary, use questions
	class motivated	use a film, or a visit to motivate

Further checks

Place—workshop, classroom—free, clean, tidy?

All equipment ready?—Handouts, summary, notes, questions Test if necessary at the end. How to use for feedback?

3

The Instructor and His Teaching

This chapter analyses some of the personal qualities necessary for a good instructor and considers how these qualities can be developed. Instructing and training men is a vocation involving knowledge, skills of many kinds, attitudes and habits of thought which can be studied and acquired just as in any other responsible occupation. Some people make better instructors than others, but all can improve if the art of instructing is regarded as a matter for study and practice, and not considered as a gift from heaven handed out at birth to some and denied to others.

The instructor has a great responsibility for laying the foundation of trainees' careers; he can start men off on the right or on the wrong foot. He is very much concerned with establishing the right human relationships, which calls for a sound knowledge of the principles of man-management. Trainees are ruthless critics of their instructors—they know at once if you are bluffing or pretending to be something you are not. They can tell from your manner and chance remarks if you have a perpetual grouse against the high-ups, or if you genuinely put your job, your ideals, or your men before any purely selfish interests. You thus infect the class with your own attitude, outlook and enthusiasm.

This is largely a chapter of dos and don'ts; of good qualities to acquire, and pitfalls to avoid. In instruction (as in everything else) it is possible to have too much of a good thing as well as too little, and we shall see that there is generally a sound middle road to be followed.

Virtues of a good instructor

1. *He must be a man of purpose.* He must know exactly what he is trying to do and must be able to persevere in doing it, so that he develops in the class a sense of achievement; this means he must know

the feel of the class and be sensitive to its needs at all times. He should always bear in mind that his job is to help his trainees to learn and to encourage them to want to learn.

Establish the right relationship with your class when you confront them for the first time by telling them your name and asking each member of the class to tell you his name. Write their names down quickly *in their seating order*, so that you have a key to the names of your class in the first five minutes of your encounter. This gives you some control of the situation—an important contact has been made and there is a psychological significance in knowing a man's name and using it. You are concerned with individuals and each one counts.

A fixed seating arrangement is recommended for all classes of more than eight as it helps in learning and remembering names. When you have to ask a question it is far better teaching manners and far more effective to call on Mr. Brown than to say 'you in the corner' or 'the big chap in the third row'. The name-key is indispensable for proper questioning of the class, as we shall see in chapter 5 which deals with the art of asking questions.

The instructor should maintain a friendly informal atmosphere in which trainees can feel relaxed and are not afraid to ask questions or discuss a point. Use verbal rewards and encouragement. 'That's a good question.' 'That's a good point—we are coming to that in a few minutes.' If someone in the class has had special experience in one or more aspects of the course, get him to contribute by saying something like: 'Smith has been dealing with this problem on his previous job. Smith, can you tell us how you overcame it?'

2. *He must know his subject and his class.* The instructor should know far more than he is expected to teach. You cannot teach up to the limit of your knowledge effectively or safely; if you try to, you will sound unconvincing and may start to bluff. In giving a demonstration, the performance must be first-class, but bear in mind that years of experience in a profession or craft do not necessarily make you a good instructor. The professor of mathematics can be quite useless as a teacher of arithmetic; the superman in a subject can easily be quite out of touch with the difficulties of ordinary men. There is further the temptation to put on a dazzling performance to an astonished, gasping class—skills and knowledge are not acquired through a mouth agape with mystification.

It is easier to show off in front of a class than to understand and grapple with its difficulties. A practical hint is to use such words as 'obvious' with great caution. It is often applied to points that are obvious only to the instructor and far from clear to the bewildered class. Its frequent use may help to bolster the ego of an instructor who is a little unsure of himself, but it has a damping effect on questions from the class. For a poor or ill-tempered instructor this may seem a desirable effect!

The importance of preparation has already been stressed, and the instructor's superior skill and knowledge do not exempt him from preparation. Whatever the qualifications and experience of the instructor there is always a plan to be followed and matter and methods to be decided on to suit the particular class. The steps of the instruction and the summaries must be thought out and a check made of all apparatus or machinery to be used to ensure that it is available and working.

When you have done all this, do not be hidebound, or a slave to a plan or the clock; be flexible enough to change your plan of campaign if you feel that it is in the interests of the class to do so. Classes are like people—they do not all react the same way. It is easy to get into a rut with instruction through experience, and even if you pride yourself that you have everything 'cut and dried', it may need more cutting, or it may be too dry! Your class is not just an assembly of units on a conveyor belt going through a process; it consists of a varied collection of individuals whom you have to help to learn.

3. *He must be enthusiastic.* Enthusiasm is infectious and can make up for a number of deficiences in a poor instructor. Over-enthusiasm, however, has its dangers and may interfere with the balanced outlook of an instructor and obscure the larger objective of the training (p. 4). So enthusiasm must be tempered with common sense, a sense of humour, and a realisation that training is a means to an end. Over enthusiasm may tend to produce a permanent teacher-dominated teaching situation with a one-way communication system.

The job of instructing is not suited to everybody. Nevertheless, a man who at first feels that it is not for him can be reasonably successful by sheer determination; but real enthusiasm produces the outstanding instructor. The class reflects the instructor's enthusiasm or his boredom. Enthusiasm is so much a matter of character and personality that specific advice about it is hard to give, but without it there is no driving force in the instruction, and no stimulation of the all-important desire to learn.

What about the ten-minute break for tea that so easily becomes twenty minutes? The enthusiasm of the instructor must be able to counteract the pull of the canteen, for both himself and the trainees.

You can and must conceal any lack of enthusiasm you may have for any particular piece of instruction that to you is uninteresting or even unnecessary. Do not expose your own lack of interest to the class. You may have heard this kind of thing said :

'This course is a bit of a bore for everybody, but we all have to go through it, and if anyone slacks I will see that he is sorry for it.'

'I am not going to stop this exercise of stripping down and assembling till someone in the class does something right for a change.'

'I haven't been able to get anything ready for you today, so I suggest you go through the calculations at the end of your books.'

Your attitude can be transmitted to your class. You may have heard something like this :

'There is a lot of talk going on about going metric. We always worked to half a thou and that should be good enough for anybody, so that's what we're going to do here. You can turn everything into metres and kilogrammes and the rest if you like, but I think it's confusing and a waste of time.'

We should move with the times and 'think metric.' With the inevitable approach of 'metrication' in all branches of technology, training officers and instructors can perform a valuable service by using the international system of units (SI units) on every possible occasion. Many industries already work in these units and appreciate the advantages of the system. It is difficult to break old habits, but very soon the pound, foot and inch will be out-moded. In schools, work is now being done in kilogrammes, metres and degrees Celcius. Many of the concepts associated with work, energy, power, quantity of heat are made more comprehensible by using the rationalised and coherent international system of units which is already well established in all electrical work.

The universal adoption of the SI units will result in a great economy in material, time and effort through standardisation, particularly in the sizes of machines, bearings, nuts, bolts, screws, building materials and paper sizes. This agreed standardisation will facilitate trade and productivity, documentation, and training. Instructors can play an important part in bringing about a change in outlook by adopting a broad-minded and flexible attitude in all instruction.

4. *The instructor should be seen and heard.* This may seem fairly obvious, but it is surprising how often an instructor puts himself in a dim light, or is obscured from a part of the class, or cannot be heard properly by all the class. A check of experienced instructors in a Service establishment showed that, for indoor instruction, about 12 per cent of the class could not see the instructor properly and about 7 per cent could not hear him. These figures go up considerably for outdoor instruction, or for practical instruction that has of necessity to be given in noisy or awkward places. Make sure that the whole class can see you and hear you comfortably, and that you and your demonstration bench or blackboard are well illuminated. Any artificial light used should shine on you and not in the faces of the class. Class comfort is important and will be dealt with more fully in chapter 4.

5. *The instructor must be able to be a good showman.* In this chapter the instructor is considered the chief character in the play. He is also simultaneously, the stage manager, and the property man, but he must know when to efface himself and become the prompter or just part of the audience. From the moment he steps in front of the class the performance is on and the instruction should move smoothly from beginning to end, with no hitches or breakdowns, either in hardware or of softwear. You can probably recall the unpleasant jarring that occurs when the projector keeps breaking down in a cinema, or when slides continually appear upside-down or in the wrong order (see chapter 10 on visual aids). The material to be shown may be good, but the effect is lost if it does not run smoothly. The instructor must have dramatic sense and a feel of the audience : he should know how to drive home his points by a well-timed use of the element of surprise, variety and an occasional good gag. Without a good sense of audience, surprises and gags can misfire through lack of spontaneity.

The instructor must ensure that his demonstration of audio-visual aids, whether films, slides, tapes or overhead projection, are of a high standard. Trainees these days are exposed to a professional standard of presentation in all kinds of media, particularly films, television and radio, and therefore anything that is second-rate or misfires in the instructor's presentation makes a poor impression by comparison and impairs the lesson and the instructor's standing. What is known as Cuss's law is always in operation: 'If anything can go wrong it will.'

6. *Dramatic sense.* Chapter 1 discussed how attention is best secured through interest, and that one way of rousing interest is to use plenty of variety, with little surprises, and an occasional touch of drama. Here are some examples.

In teaching mechanics there are numerous pieces of equipment that can effectively be produced to save the lesson from too much talk and chalk. An ordinary bicycle brought into the classroom can form the basic apparatus for explaining mechanical advantage, velocity ratio, and levers.

The element of surprise can be used when explaining horsepower. Invite one of the trainees, preferably an athletic type, to run up two or three flights of stairs as fast as he can, and time him by means of a stop-watch. Let the class look on and get them to measure the height of the stairs and weigh the athlete. Then return to the seclusion of the classroom and work out the volunteer's horsepower during this effort.

This affords an opportunity to use SI units mentioned earlier on p. 33. Suppose your athletic trainee weighs 75 kilogramme-force—in SI units of force this is $75 \times 9{\cdot}81$ newtons; the height of the stairs is 6 metres and the time was 5 seconds.

The work done by the trainee (in joules) is calculated as the force overcome (in newtons) × the distance (in metres).

$$\text{Work done} = 75 \times 9{\cdot}81 \times 6 \text{ joules}$$

$$\text{Work done per second} = \frac{75 \times 9{\cdot}81 \times 6}{5} \text{ watts}$$

$$= 885 \text{ watts}$$

(He was working about as hard as a single-bar electric fire.)

But 746 watts is one horse-power. So, for a few seconds, the trainee worked at an average power of

$$\frac{885}{746} = 1{\cdot}2 \text{ horse-power}$$

This break away from a formal class routine provides class activity, and gives the class a practical idea about horse-power which they will not forget.

In a course on safety precautions or first aid, arrange for someone to come into the room in some excitement and announce: 'Old Thomson has had an accident. He is lying on the floor and looks as though he has passed out.' Use an actual prearranged situation to test how they would deal with the emergency in a practical way—a judicious use of tomato sauce or red ink will add a little realism. Some audio-visual aid firms have produced very realistic simulated wounds or burns that appear quite gruesome when applied to a simulated casualty. People undergoing first-aid courses must get used to the sight of wounds. Let the class work in pairs, and then discuss what each pair did well and what they did badly.

This approach could have a number of applications, not only to provide practice in first aid. Some simple role acting, with trainees taking part, might be devised to demonstrate the importance of safety precautions; for example, the danger of leaving things lying about on the floor could be demonstrated, or not switching off the current and drawing fuses before starting any eletrical repair work.

Safety precautions and dealing with emergencies are techniques that can only be learnt by *doing* and practice. You cannot learn a technique fully by studying a book, or listening to people talking. Make the instruction practical. Rouse interest by surprise, variety and a little good dramatisation. There must be participation.

Fire precautions and fire-fighting offer plenty of scope for surprises and dramatisation. It is well worth letting off a few fire-extinguishers or smoke-bombs and having to clear up the mess if the full objective of the course is satisfactorily achieved. The test is; does it help the trainees to learn?

7. *Timing.* Many people have a special facility for communicating with an audience, or a class, but this facility can be developed. Actors, and orators are skilled in their ability to 'get across' to their audience and hold their attention. Much of their skill depends on a sense of timing derived from watching the mood of the people in front and reacting to the signals emitted by their expressions, facial or verbal.

A good piece of instruction, like a good sermon, should have an object, a good beginning, a few points, and the 'message' should be punched home at the end. Start on time and finish on time. It is doubtful whether anything worth while is achieved if you keep your class past the time scheduled to stop, particularly if you encroach on the dinner hour or there is a clock facing the class. Impatience and restlessness set in and prevent you from making an impressive ending and punching home the points you set out to make.

A proper sense of timing can be developed only by experience. An actor knows just how long to wait for the applause or laughter to die down so that nothing is lost to the audience before going on.

In a similar way, the instructor gets the feel of his class. Any new and arresting fact should be followed by a very reduced tempo to allow time for this new idea to turn over in the minds of the class. It would be ineffective instruction to follow on immediately with a rush of further important points.

Vary your methods of approach to a difficult point and slow right down till you are reasonably sure that the class is with you. Watch the faces. They are the dials of the machine you are in charge of; they indicate whether your speed is too fast or too slow. Encourage trainees to participate, to ask questions and to discuss. This helps them to learn and gives you feedback.

The instructor is in some ways like a conductor of an orchestra. He does not appear to make any audible contribution to the output of the orchestra, but he gets the best out of it, ensures co-operation, and helps each member to give of his or her best. He works harder perhaps than any other member, as is evident by the way he mops his brow after each session.

8. *Manner.* The term 'manner' is here meant to include how an instructor stands, moves about, speaks, uses his hands and eyes, and his expression. These are all part of communication and will be dealt with under the headings : voice, eyes, mannerisms, stance, attitude. The instructor is the most important audio-visual aid to communication and learning.

Voice. Speak loud enough to be heard easily at the back of the room, but do not shout. Speak clearly and distinctly, using your natural voice, and avoid any temptation to imitate the style of your favourite film star or politician. Speak slowly, particularly if you have a large class, but vary your speed, emphasis and tone from time to time to avoid becoming monotonous. Avoid the bad habit some speakers have of allowing their voices to drop at the end of a sentence. We are very poor judges of our own way of speaking, so arrange to have one of your lessons taped. Play it back after a few days and then criticise yourself ruthlessly. A good way to root out faults is by mutual assessment of instruction by other instructors (see p. 49).

If you are prone to run your words and sentences together too much, try to make yourself pause after each sentence for a good two seconds, and for one second after any important word. Spacing speech requires practice, which can best be done in private. Do not ever be afraid to pause if you get tangled up in your words. A deliberate pause passes unnoticed and produces a little expectancy, but a series of 'ers' or 'ums' to fill in time advertises that you are stuck for words.

Eyes. Look at your class, let your gaze travel round the class from time to time; but it is useful to have a point back-stage centre, a few feet above the heads of your class, as a base from which to operate and to return to. In this way you can make each man in the class feel that you are speaking to him individually, and you have the whole class fairly well covered. There is real power in the human eye, so make good use of it. Avoid delivering a lecture to the vault of heaven or to the window at the side of the class, or to the blackboard.

Mannerisms. Avoid doing anything which will distract the class from what you are saying or demonstrating. Very few instructors are free from mannerisms, but the particular way a man stands or moves about or gesticulates does not become a matter of any great importance unless it distracts the attention of the class.

Mannerisms are often nervous habits, such as throwing a piece of chalk a few inches in the air and catching it, tapping the side of ones leg with the pointer, or fidgeting with something on the bench. Remove all temptations to fidget. Coins or keys in pockets somehow lure lecturers to jingle them. Do not cling to the rostrum, desk or chair. You may feel as if you want to be rescued, but don't show it!

Nervous verbal habits are also distracting. A common mannerism I have noticed in some Service instructors is to punctuate practically every remark they make by barking in a challenging way the query 'right?' followed by a stern gaze over the class as much as to say: 'disagree at your peril.' Very few men under instruction of this sort feel it worth while to shout out 'wrong' or 'I am not following you'. It is a good idea to find out at all stages whether or not the class is with you, but working the query 'right?' to death defeats its own object. Better methods are suggested in the section on questioning (chapter 5). There are other verbal bad habits that are used over and over again as a sort of padding. They become outworn tags that have no particular meaning for the class and when let loose continuously can have the same effect as a sleeping-draught. Some students have been known to keep themselves awake by maintaining a score of the number of times a favourite tag is used, and running a sweepstake on the result.

A common distracting mannerism is ''umming and 'erring'; already mentioned under the sub-heading *Voice.* The speaker may perhaps feel that the practice creates the impression of making great mental effort, but the habit is most distracting.

Stance. Unless you are a good actor, and know all the tricks of standing and gesticulating, the best advice is to stand easily on both feet, without being rigid or sloppy. Do not sit (unless conducting an informal discussion or talking to a small group), lean or slouch.

We communicate with each other a great deal by gestures and facial expressions, and these are to be encouraged as long as they assist you in putting over your thoughts and feelings. If, however, you find your hands merely wanting to fidget, put one behind your back and let the other hang by your side until called on to support your words with an appropriate gesture.

Your appearance can be a distraction to some if your dress is very untidy, or too flashy. Instructors are often very critical in judging their students, but in their complacency they do not always realise that students apply a far severer and stricter standard in judging their instructors. The instructor himself should be a vital audio-visual learning aid.

Attitude. Finally in this catalogue of virtues we have to consider the attitude of the instructor to his class. The establishment of a good relationship between instructor and class is vitally important, and determines whether the process of learning is going to be a co-operative effort, an uneasy alliance, or a cold war. Army instructors are given excellent advice in a brief maxim : 'Your attitude must be *fair, firm* and *friendly.*' The instructor's fair mindedness should spring from his own personal standards of honesty and integrity; his firmness from a sense of control and discipline in his own life; and his friendliness from unselfishness and a courteous consideration for others. He must be sympathetic and approachable, both over a student's work, or personal problems. These are essential qualities in any well-integrated personality; they also sum up the essentials for good management.

These may seem formidable standards to achieve, but the instructor must set himself this high standard in dealing with his class, and then practise making the most of his gifts and personality.

The well-known lines of Burns :

> 'O wad some power the giftie gie us,
> To see oursel's as ithers see us,'

are particularly apt for the instructor. It would do a great many instructors good to be able to listen to the remarks of their students when they are discussing them after the instruction, as they invariably do.

This idea was elaborated in a Services course for instructors, and provides an example of the 'role playing' way of introducing more active, class-participation methods of learning in place of the formal lecture. As an introduction to a series of small discussion groups on the essential qualities of a good instructor, the class listened to what was purported to be a taped record of various snatches of conversation picked up by a hidden microphone in the canteen after a morning's instruction. The rights and wrongs of all concerned in these conversations provided points for discussion. Here again the use of the tape-recorder as an audio aid can be demonstrated.

Here are some examples that might be used during a course for instructors. B (for example Bloggins) in each example is the fictitious instructor—but not the same one in each case.

'That question of Taffy's fairly shook him.'

'Yes, didn't it? He did his best to talk himself out of it, and laugh it off, but you could see that Taffy knew what he was talking about, and B was just bluffing.'

The moral here is never bluff. Know your stuff, and if you don't know all the answers (and very few people do), say you don't know but that you will find out for the next lesson. Then, of course, keep your promise. You will not lose face by being honest, but you will by bluffing.

'I heard that yarn he told us when in primary school.'

'He's always trying to be funny. I suppose he expects us to laugh !'

The point here is, do not throw away your position as an instructor by trying to curry favour with the class. Familiarity breeds loss of respect if not actual contempt, and some jokes or wisecracks are not appreciated by all members of a class.

'I didn't expect the big shot to come barging in and take over like that, did you?'

'I suppose he has got to see what is going on and show us he is the boss, but B took a poor view of it I think.'

The relationship between the instructor and his class is sacred, and should not be interrupted. Instructors should never be disparaged in the presence of their class, nor behind their backs. Instructors in an establishment must be absolutely loyal to each other, to those above them, and those below.

'B always seems to pick on Nobby to answer all the questions or to dance out in front.'

'He knows his old man and that's why he put Nobby on as a special turn when the governor came along.'

'Nobby is the blue-eyed boy all right.'

Every member of the class is entitled to a square deal. You must be fair to your class in every way. Favouritism, or any kind of unfairness, loses you respect. Do not try to shift the blame for any of your mistakes or weaknesses on to the class.

'We started B off again on his R.A.F. days. He can't keep off it.'

'I am glad he went off at a tangent again as I hadn't looked up a thing about machine tools. I bet we can fix it tomorrow too.'

This snatch of conversation shows the importance of being purposeful, and knowing your own mind. Do not follow red herrings and never allow slackness or slovenliness to pass unchecked.

'He knows his stuff all right and knows how to put it across, but he somehow gives me a pain in the neck.'

'He always tries to make you feel small.'

'I'd like to see someone cut him down to size.'

Being firm and insisting on things being done the way they ought to be done does not mean being remote and unapproachable or a petty dictator. With true insight into human nature, an instructor can gain the respect of the class in a friendly atmosphere, even though he has to pull men up for faults or slackness. Guard against any tendency on the part of the class or yourself to make a man look a fool when he appears to be particularly stupid, or to possess 'two left hands'. A timely comment from you will help, such as: 'You may think it amusing watching Hamhand doing this, but he can beat you all when it comes to working out the formulae.'

'B ought to have his act *The new man starting on the lathe* put on the television. It's the funniest thing I have seen.'

'He keeps such a straight face himself, and doesn't look a bit as though he is trying to be funny.'

The timely use of a little humour, including the very occasional 'wrong way' demonstration with exaggerated faults ending in disastrous results, does a lot to establish good relations between instructor and class. It releases any tension, lubricates any points of friction that may exist, and can often be used to focus attention on an important point. It can rally the class from a comatose state of semi-consciousness. But the joke must come off! The class has little sympathy for an instructor–comedian whose jokes fall flat. Even when well done, there can be too much of a good thing. 'Wrong way' demonstrations exaggerated to be funny should not be turned into a 'Crazy Gang' show more suited to a music-hall.

Beware of earning a reputation for being the funny man. You will often secure the attention of the class, because it is waiting for the next wisecrack, but the instructional value of your lesson may suffer. Any humour should be brought in casually, as a part of the instructor's way of expressing himself. It is impossible to lay down any rules on this subject. Humour is not essential to good instruction; it is useful and is rather in the nature of a luxury. Be yourself and if humour does not come to you naturally, do not pursue it. The test is does it help the trainee to learn?

'Dusty Miller got it in the neck today, didn't he?'

'What was it B said to him? "If your brains, Miller, were to turn into gunpowder, they wouldn't blow your hat off." Poor old Dusty.'

'If he had said that to me, I should have had a showdown.'

Sarcasm sometimes gets a cheap laugh, but it is rarely worth it. If you make a man look a fool he will find it very hard to forgive you or forget it. Show a calm unruffled front to the class at all times. You achieve little or nothing by becoming exasperated, except to provide a little secret amusement for the class. *Do not nag.*

'What was it B wanted you for this morning?'

'He fixed up a time on Saturday morning to give me a run-through on starters and controllers, because I messed up my test.'

'On Saturday morning? Old B's O.K.'

A man appreciates any sort of personal attention you can give him, and is stimulated by the knowledge that you know his good points as well as his bad ones. Know your class and show that you know them all. For example :

'Here, Rogers, you are the expert on turning, come and set this lathe up for the job shown on the card.'

'Now, Bell, you came out top in wiring. I want you to be in charge of the party for re-wiring the lecture room. See me tomorrow at 10 a.m. about drawing the stores for the job.'

Take note, too, of the bottom third of the class. An extra class or two for the less bright will help you to find out just where they get stuck, and give them a chance to catch up with the others at their own speed without holding back the rest of the class.

'I have never yet been able to do anything right for B. Nothing anybody does is good enough for him.'

'I suppose he harps to every class he has, that they are the worst bunch of idiots he's ever come across.'

There is nothing to be gained from making a class feel that they are the worst lot that has ever taken the course. If things aren't going right, do something positive about it. The best plan is to take stock of yourself. An instructor who is forever pointing an accusing finger at his class, and blaming them for all the inattention, bad results, or disorder, should note that while his index finger points at the class, his three other fingers point back at himself! A critical nagging attitude will irritate the class, and good relations and co-operation will be impossible. It has been found that the best results are obtained by a good deal more praise than blame. Praise should be handed out with discrimination, and to particular individuals rather than *en bloc*. 'Well done,' or 'Well done, that man in the back row,' will not be as effective as 'Well done, Smith.' It is, as we have seen, important to learn the names of your class at once and address a man by his name.

Dealing out praise is more effective when it is given a little *crescendo* effect, such as,

(1st time('There is an improvement there, Taylor';
(2nd time) 'You are doing well now, Taylor, keep it up';
(3rd time) 'Taylor, that was a first-class piece of work.'

One often hears a seasoned instructor telling a class, 'This is dead easy,' 'There's nothing to it.' This may be so for the instructor but it may be quite different for the class, who become discouraged. It is better psychology to be quite honest and say, 'This is not at all easy and is going to be a severe test.' Such remarks serve as a challenge and bring out the best in the trainees.

Additional points concerning class management or discipline

Discipline is a word often associated with the rigid autocratic control of people, but for the classroom and workshop it is the maintaining of prescribed standards of conduct essential for efficient teaching, class participation, good personal relationships and sound learning. At the end of this chapter are some hints on how to establish good attitudes and an atmosphere of orderliness in the workshop or training area. They are well known to most instructors, and are important in achieving the larger objective of training referred to in Chapters 1 and 2.

The instructor achieves the most effective kind of discipline by personal example—not by preaching or exhortation. Each kind of training or work has its own approach, depending on whether the activity is a lesson, discussion, conference, laboratory work or field work. It is not sufficient for the instructor to see the work as important, interesting and worthwhile; he must also make his enthusiasm infectious. Many of the valuable results of training are not taught, but caught.

You should keep rules to a minimum, and never make one that cannot be enforced should the need arise. The purpose of rules should be understood and accepted by all as being necessary for the ultimate benefit of the whole class. 'No smoking' is a rule in most training areas. See that it is observed.

Keep the class occupied and interested, and have ideas and minor projects prepared for occasions when the normal plans for the day go wrong. Orderliness in a group of trainees should emerge because they are too busy and interested to have the time or the desire to cause trouble.

Disorder and poor discipline may arise through poor communication leading to misunderstanding. An order or instruction given verbally to a student can take on several shades of meaning: What you *think* you said and how you said it; what you *did* say and how you said it; what the student *thought* you said and how you said it. Instructions must be clear. An instructor who has difficulty in keeping order should look to his own methods and manner.

Frequently in technical colleges there are students who are under some compulsion to attend, and have no real inclination or desire to learn. To be able to win their interest and respect is a real tribute and calls for much skill, tact and understanding.

Instructors and teachers who have least trouble with their classes make it known from the start just what they expect by way of behaviour. In the past, long lists of rules were drawn up, which were intended to regulate conduct in class, and included points on punctuality, horseplay, chewing, smoking, courtesy. Such rules are not so much in evidence today, but it is as well to let it be known that you expect at all times a high standard of courtesy and good classroom manners. Stamp on deviations firmly, fairly and courteously as they arise, on the grounds that nothing works in this world, or its institutions or groups, without consideration for others.

In classes of 20 or more, you will always find the talkative, inattentive types. Such offenders can often be dealt with by pausing unexpectedly and abruptly and fixing the main offender with a stern glare—this usually produces the desired effect. You might then politely say 'Thank you,' and carry on—or just carry on. Your instructions or commands should be authoratative, clear and brief. It is useless to say, 'Stop talking' every few minutes; an overworked command to all the class is quickly ignored—as it is taken to apply to no one in particular.

It's a good rule to direct any rebuke you have to a specific person by name, in an imperative tone of voice, but on occasion a touch of humour will go a long way to restoring good order and good relations.

Be consistent in the standard you expect—it should be independent of the time of day, or your own mood or feelings.

Avoid anything which is likely to blow up into an embarrassing situation in which your authority is in head-on conflict with the students—or a particular student who is seeking a showdown.

Never punish an honest mistake and never take drastic disciplinary action without full and careful consideration, and then do it calmly and firmly. In the last resort, a real rebel has to be isolated—but make sure you have the group behind you :

'We are here to learn and we don't want to waste our valuable time. We all have to work hard to get on top of this subject. So it's quite simple—you, Smith, either behave yourself, or leave the class.'

The essential point is that the respect and opinion of the group is behind you, and not behind the offender—who will show off to gain the approval of the rest of the group.

A useful move is politely to invite the awkward student to see you after the class. You then have a chance to ask him what is wrong, and have a confidential talk. You may discover many causes for refractory behaviour : home situations, something you said days ago; other students, trouble with the police. Whether you have to show sympathy and understanding, or administer a rebuke, or a bit of both, depends on all the circumstances—but it has to be carefully handled and away from the group. The important thing above all others, is to show that you are interested in his welfare and his future, that you respect him as a person, and that you are there to help him in every way. It is worth trying to give him some responsibility, and get to know more about him in a friendly way.

It is a very sound piece of advice in dealing with people to put yourself in the position of the other person, and try to see through his eyes. If you can do this honestly, your attitude will be free from any serious faults. Do not imagine that the points, so lengthily discussed here, can be applied like a magic formula to every class or situation you are likely to meet; allowances and modifications have to be made to suit each particular class and each subject. A class of young apprentices on an induction course needs different handling from that of senior technicians learning a new works process.

Women trainees

Today women employees are an important factor both numerically and qualitatively in our work force. The instructor may be a good judge of men, but he may be less effective in dealing with women. We have to start this section with a few basic facts about women in industry as trainees.

A woman can become just as skilled as a man on a job and are as intelligent as men. This is amply borne out by the wide variety of skilled jobs being done by women in USSR.

Compared with men, women tend to react to situations involving people in a more personal way; they are more interested in people than in abstract ideas. Women, in general, are less able to lose themselves in their work, and are usually more dependent on praise and encouragement. As a rule, women are happier when working with others, and conversation with others and companionship at work is important. Women tend to react more emotionally to criticism.

A basic principle of any supervision is that criticism or censure of a person should be given in private away from his other companions, and in sober judgment. Women are often more conscientious in their work. Their physical ability is less than that of men, but there are wide variations among men and women. We find that women are just as good, if not better, than men in withstanding fatigue. Their powers of endurance are remarkable, and they can achieve the same work production efficiency and quality standards as men.

Finally, nowhere are the personal standards of conduct of male instructor or supervisor more exposed to criticism or more influential on the attitudes and behaviour of trainees than in his dealings with women trainees or workers. There is a time and a place for everything, even flirting, but working hours and the training area or workshop are not among these. This should be an absolute rule to which all must be committed.

Further points on class management are dealt with at the end of chapter4 and in chapter 7 under 'Group structure'.

In dealing with the qualities required for a good instructor, it is a good idea, during a course for instructors, to invite the course members to analyse and make a list of all the qualities they would expect to find in the perfect instructor. The resulting synthesis might be a bit of a prig, but the exercise might prompt useful discussion on what could happen if you have too much, or too little, of a good thing.

The conclusions might end up looking something like the following Table, on the basis of a well-known soap-advertisement concerning its lather: 'not too much, not too little, but just right'—there is much to be said for the 'golden mean' of the Greek philosophers.

One of the important aspects of the work of training instructors is

THE QUALITIES OF A GOOD INSTRUCTOR

An instructor	*Not too little—*	*Not too much—*	*But just right*
be purposeful	no clear objects wanders no sense of time	enslaved to a plan enslaved by the clock and timetable no flexibility	object clear keeps to the subject but deals with interesting points keeps an eye on the clock
know his subject	skills—poor demonstration knowledge—only a little ahead of class techniques—dare not leave the text-book	shows off dazzles the class with science becomes the talkative veteran	skilled demonstrations knows his stuff instruction backed by practical experience
be painstaking	lazy in preparation haphazard and muddled in methods	unadaptable too attentive to details pernickety	good preparation sense of balance and proportion
have dramatic sense	stodgy dull monotonous	spectacular performance but object lost in the spectacle too much variety	uses dramatic sense in proper place introduces variety
be enthusiastic	apathetic uninterested rejects new ideas	fanatical unbalanced in outlook, to the point of being boring	keen infects the class with his enthusiasm
have pleasing manners	fidgety, nervous, jarring voice awkward gestures	smooth, oily smarmy slick	at ease with class confident, pleasant voice, eyes, natural mannerisms, *stance*
have control of class (attitude)	has favourites no control too familiar soft 'this is a rotten job but someone has to do it'	undiscriminating domineering never relaxes sarcastic 'the works are run for the benefit of instructors'	fair firm friendly respected—sets an example sees through the eyes of the class

to develop essential personal qualities. Knowledge of the subject is only one essential; at least half the time should be devoted to the improvement of an instructor's manner, attitude and outlook. In a very successful Service school for training junior instructors, more than 75 per cent of the periods forming the course were spent in actual *practice* in instruction.

An instructor's performance in front of the class, including all the qualities just discussed, can best be assessed by one or more fellow-instructors. Technical teachers in a works or training establishment tend to work in isolation, and rarely have the opportunity of seeing other teachers at work or discussing teaching techniques and problems.

Much could be done to improve the quality of instruction, without rousing the resentment normally associated with criticism, if instructors and training officers could hold a seminar among themselves on the techniques of technical training—using, for example, some sections of this book as a basis for discussion. An outside expert could be invited to act as chairman or leader of the discussion. Generally people do not find it easy to assess another person's ability to teach, but the task becomes more manageable if each aspect of the instructor's job is analysed separately.

A suggested list of the points to look for in a piece of instruction is shown on p. 50. The assessor, or fellow course members of a training course for instructors, should fill in the applicable parts of the form at the end of the session. Filling in the form during the lesson is not recommended. A course for instructors should provide several opportunities for each course member to give lessons, trade talks and practical demonstrations to his fellow course members, or to an actual class, so that his performance can be assessed and discussed in a helpful and friendly way. This process is called microteaching.

A great asset to teachers and instructor trainees is the use of the video tape. It is now fairly simple and inexpensive to take a video tape of an instructor conducting a lesson as a demonstration, and to show it to him afterwards. This can be done in private immediately after a performance, so that he can criticise himself. The tape can then be shown to his class mates and the instructor conducting the lesson can say what he thought went wrong or could be improved, and the class can offer their comments, followed by a discussion and comments from the course leader. This is proving to be an excellent way of providing the gift besought by Robert Burns, p. 40.

The list of points to look for is given in this chapter as it deals with the instructor and his performance, but the actual use of the form as an assessment guide should be deferred until after a study of Chapters 4 and 5 dealing with the class, the learning process, communication and assimilation. The form should not be used as a yardstick for measuring instructors, as many personality factors have to be considered to do this at all satisfactory, but use the form rather as a guide to help them in their job. As we shall see in chapter 6, assessment properly used is an important aid to learning.

Here is a suggested check list which could be used as a guide for assessing instructors. Section I deals with the *teaching* situation; Section II with the *learning* situation. The latter provides the ultimate criterion for judging the success of a piece of instruction.

The following marking on a five-point scale is suggested :

A	yes
B	on the whole yes
C	rather doubtful
D	on the whole no
E	no
or N.A.	not applicable

I Did the instructor : Mark

1. Put the class at ease at the start?

2. Know the names of his class, or have a seating plan?

3. Stimulate interest initially. Introduce the subject well and start with things they knew?

4. Have a clear *objective* and state it or define the job clearly?

5. Keep the objective in mind and go for it purposefully and without digressing?

6. Check the trainees' existing knowledge?

7. Know his subject well? *Prepare* his trainees, his lesson, his demonstration?

8. Ensure the work place was tidy, and free from obstructions or distractions?

9. Plan the session well, with good sequences and summaries?

10. Ask well-prepared questions to make the class think and participate? Encourage questions?

11. Ask questions to confirm learning or ensure understanding, and deal with the answers skilfully?

12. Set apart adequate time for closing the session in an orderly manner with equipment put away and the workshop tidy?

13. Stress essential points; i.e., what the trainees *must* know?

14. Prepare good audio-visual aids to communication and use them effectively?

15. Achieve a good *presentation,* making the steps and stages clear?

16. For a skill: Make a careful analysis of the skill with proper attention to the key points, including safety and detailed steps?

17. Present the operation with trainees suitably positioned?

18. Have an effective build-up of the subject with a summary for note taking?

19. Attempt just enough for the trainees to master in the time?

20. Include adequate detail—without haste?

21. Assess the work or effort of each individual as far as possible?

22. Let him know how he is getting on?

23. Tell him how to improve?

24. Give proper praise whenever deserved?

25. Give any reproofs in a fair and helpful way?

26. Have a good manner before his class?

27. Make himself heard and understood?

28. Maintain good *communication* with his class? Watch their faces?

29. Have a good stance, free from distracting mannerisms?

30. Have good control of his class?

31. Have an attitude which was fair, firm and friendly?

II Did the class:

1. Settle down quickly and in good order?

2. Show interest and keenness throughout the session?

3. Participate in the lesson and have to think in a learning situation?

4. Ask questions?

5. Appear to understand and *assimilate* the new knowledge or skill to the required standard?

6. Get suitable and useful notes; e.g., from own observation, audio-visual aid or hand outs?

7. For a 'skill': Have adequate time to practise the job correctly to the standard required for the session?

8. At the end of the session show signs of wanting to discuss points?

9. In general appear to work well together as a group, and appear to appreciate and co-operate in the character training aspects of Section I, such as in 8, 12 and 31.

Notes on instructional management for instructors, concerning the larger objective of training discussed in chapter 1, which dealt with acquiring good habits of thought and work.

Care of machines and tools

1. You are responsible for the condition of the tools and machines in your shop or classroom. As a careful worker, you well understand the importance of keeping them clean and in good repair, and accounted for.

All machines and tools should be cleaned after each period of use. In addition, most machines need special care, and regular lubrication.

All machines should be tested and adjusted periodically for bearing wear, alignment, safety of electrical and mechanical devices, gear wear, and other common mechanical defects.

2. Each trainee should be given clear responsibilities for the proper maintenance of his machine, bench or work station down to the floor. The average trainee takes this work in his stride as part of his job, if he is told about it when he enters the shop. To encourage responsibility—give responsibility.

Storage—a place for everything and everything in its place

3. Bar steel and sheet aluminium should be stored in well-designed racks. Bolts, nuts, rivets, and other small items should be kept in appropriate bins or boxes. Glass supplies should be put away carefully to prevent breakage. Keep in mind safety, protection from damage, as well as orderliness and convenience of use.

4. Storing hand tools requires adequate space on wall panels, shelves, racks, and in central tool cribs and cabinets. Use the facilities that are provided, and, if necessary, improvise others in any available space. Hooks may be used to hold many small tools in place. Drills, bits, chisels, and the like, fit in holes on special tool

boards. Machine accessories and special tools may be stored in the main tool crib or in separate places near the machine.

5. A special place should be provided for each tool. If you draw the outline of the tool as a background around its hook or place on the tool panel, you can inspect the tools easily for loss or breakage at the end of each session or day. To help the trainees learn the names of tools, label each tool or its place. You can mark the names on the tools by stamping, etching or painting.

Storing the finished and unfinished work of the trainees is required in most workshops. Usually, small equipment or accessory jobs, such as electrical instruments, should be safely put away on stands, in bins, racks, drawers or cabinets.

6. The personal belongings of the trainees need to be cared for in lockers or cabinets which can be locked. Since many of the trainees may have their own small tools, space for their storage should be provided.

7. It is particularly important that a trainee should immediately become acquainted with the tool and accessory accounting system.

To keep the trainees alert about safety

8. Good workmen are not careless. Trainees whose safety education accompanies their daily work in the shop get into the habit of working carefully. In addition, an instructor should seize every opportunity to make them aware of the importance of safety practices.

9. Keep workshops clear of all 'junk' and everything that makes the place look untidy or could cause accidents; they should be ready for inspection by the highest authority at any time.

Closing the class period

10. This should be planned so that an efficient time-saving routine is established. Without some routine, trainees awaiting the close of the period or dismissal of the class tend to waste time and become disorderly, and even to leave the workshop untidy so that the instructor has to spend time putting things away.

11. For a practical class, each trainee should clearly understand what he is expected to do; he should also understand that he is expected to work steadily on at his job until the end of the period.

A proper length of time (not *too* long, that time is wasted) should be set aside at the end of each practical period for cleaning up and putting tools and materials away.

Any tendency to idleness, on completion of a particular task, should be counteracted by the instructor having a list of short jobs or duties to be done in connection with the efficient maintenance and operation of the workshop, which can be assigned at such times. These duties should be useful activities which need doing and which have instructional value.

12. Workshop routines differ, but in general time should be allowed for the following :

i. all tools, small equipment, books, instruction sheets and other items used by trainees should be turned in, checked or put properly away;

ii. a brief check should be made that all trainees are present—this could be important in the case of an accident or other emergency;

iii. trainees should then be ready for dismissal at the end of the period, and leave in good order.

13. Always keep in mind the long-term objective of training—as well as your immediate objective.

4

The Class and its Learning

The learning process

In this chapter technical training is considered not as a performance by the instructor acting as a transmitting station, but as an activity of the class at the receiving end of the process. The instructor should stimulate a full contribution from the class, and obtain its co-operation in setting up an interchange of ideas and experience. Teaching is essentially a matter of communication, not simply transmitting.

Nevertheless, no matter how skilled the instructor may be, learning is essentially the task of the learner, and can be accomplished only through the effort of the learner, who must be in a responsive frame of mind. He must have the *ability* to learn, which is associated with intelligence, and also the *desire* to learn. If a man does not *want* to learn, you can use every known pedagogical artifice on him and he will learn nothing. A bad instructor working on a class which wants to learn will achieve far more than a good instructor with a class which does not want to learn.

The instructor's responsibility is to see that all the motivating tactics available are used. One of the hard lessons an instructor has to learn is that, despite the prominence given to the instructor in chapter 3, his main job is to create and work in a *learning* situation rather than in a *teaching* situation. Good instruction should be *trainee* centred not *instructor* centred.

Interest is essential for attention, and attention is essential for learning and memory. Interest here does not mean simply a localised interest in a particular part of the syllabus, *or a day's work*; such a fragmentary interest may soon wane. The interest that *motivates* on a long course is something deeper and is based on fundamental emotions and desires. For example, interest of the sort that motivates apprentices is roused, encouraged and maintained by a number of basic urges or desires. Some of these are listed below and opposite each is a suggestion for using this desire as a motivator.

Basic desire	*How to serve as a motivator*
1. desire to qualify for higher grades or diplomas desire to earn more pay desire to 'get on' for the sake of a family or fiancee	show the purpose of the instruction, its worthwhileness, its importance to the trainee personally, and its relevance to examinations
2. desire to 'make the grade' as a matter of self-esteem to stand well in the esteem of others	progress-reporting, and by a proper use of rivalry and competition in groups give individual attention
3. desire of the artisan to do a job well, and to take a pride in his work	use all opportunities to illustrate and display work of special quality and craftsmanship a proper distribution of praise and rewards the teacher must be respected for this to be effective
4. desire for the satisfaction of curiosity and of intellectual achievement, which makes a worth-while subject intrinsically interesting	encourage the learners' curiosity and make use of the elements of realism and surprise use variety give opportunities for discovering things for themselves, and to look things up; e.g. use the library use the discovery method
5. desire to express oneself and to take an active part in the activities of a group	keep the class as active as possible use project methods, role playing and discussion methods with planned questions and answers give responsibilities
6. natural desire to be as comfortable as one reasonably can when undergoing instruction	give proper attention to the physical comfort of the class and class morale generally

The motivating devices on the right require further consideration.

1. *The learner should be made aware of the importance of the learning to himself and his future.* There should be a definite purpose behind every piece of instruction given. The class is reasonable, and if you reveal this purpose it will make the instruction more relevant to the course, and personally applicable to the ambitions and aspirations of each member of the class.

The class will regard you as a person who is putting them on the right track and helping them to qualify. Here we see the importance of your preparation work in sorting out your subject matter into *must* know, *should* know and *could* know. A trainee in the midst of a hard course is right in not exerting himself over the things that he believes to be unnecessary frills. Your guidance here and explanation of the purpose of it all will be valuable.

Examinations and the award of diplomas are useful motivators. Most apprentices and men on a course are fully aware of the importance of getting on in their trade or profession and each may prefer to keep this thought as a private spur for himself. Encourage men to look to their future, but do not exclude the other ways of motivating.

2. *The use of competition and co-operation as motivators.* It is inborn in us to be competitive. We are aware of this in the widespread popularity of sport, and in thousands of other ways from the way we dress to the way we swap stories. It has been estimated that nine tenths of the world's work is done through rivalry. Our evolution has been dependent on competition with our environment and with each other for survival; but equally important and indeed necessary for survival has been, and still is, the inborn urge towards co-operation between individuals and groups.

Competition tends to produce greater output, and members of competing groups learn skills faster. The *quality* of work done by groups in competition, however, suffers in comparison with groups not trained under conditions of competition.

One of the important objectives in training is to produce people who are co-operative and get on with their colleagues in a spirit of loyalty and mutual understanding based on good communication. It has been found that groups motivated to co-operate are in certain circumstances more efficient than groups motivated entirely by competition.

The project method of training, and also the discussion method mentioned in chapter 8, successfully use co-operation as a motivator. Participants in such methods make their own contribution to a co-operative effort and have a sense of responsibility and belonging.

There are thus differences in outlook on the use of competition in training programmes. Competition and also co-operation can be regarded as both teaching methods and learning aids. They should provide for all concerned enjoyable motivation and participation. If they do not help trainees to learn in any particular situation, they should not be used.

Competition therefore can have an important part to play. We all have to stand up for ourselves, and we like to get a good place in the class. The competitive spirit can, with moderation, be successfully used, particularly between rival groups. For example, suppose that on a course on Works Safety a fire-drill has been worked out, and it takes one group 2½ minutes to carry out the emergency procedures. A good instructor could encourage useful competition between rival groups by using a stop watch—setting one team to improve on the performance of the other. You thus have keen competition between the groups, but good co-operation and mutual help within each group. Here you have both competition and co-operation acting together as motivators.

Where time and productivity are important, well-planned competition in training on actual processes can be very effectively used. Competitions between groups in first aid, workmanship and handling materials should be carefully organised—a few spectators will do no harm if the show is good.

A little group rivalry in training can also be used in ordinary class instruction. One way is to divide the class into two groups: get each half to prepare, say, a dozen questions, which they fire at the other in turn; allot marks. In this way keenness and interest is developed through the competition of the opposing halves. Stirring up one half of the class against the other, or one group against another, in friendly rivalry is more desirable than competitions between individuals, which should be avoided—it tends to make the bright boys, who always come out on top, feel cocksure and the not-so-bright feel rather depressed.

Rivalry is ruined if there is too great a difference in performance between those competing. A football match where the score is 10–nil lacks interest and spirit. Vary the competitions so that everyone has a chance to do well at something and the opportunity to co-operate with others as a team.

3. *Progress must be known.* Men who are training are all, in a sense, competing, if not with each other, then with a standard of achievement. The trainee must have a feeling of progress and should not be allowed to slack off as soon as some particular training objective has been achieved.

When the class has reached one goal, set another, so that there is a feeling of continuous effort and progress. Nothing is more deadly to a class than the feeling that they are marking time. (Sometimes this feeling is there when in actual fact the instructor is aware of definite progress.) The class must be made to feel that it is making progress, and it is the job of the instructor to encourage that feeling.

It may happen that a particular person's progress is slow and heavy going, giving him the feeling that he is never going to get through the course. This is a common problem and the line to take is to make a small mountain out of any molehill of progress and to show it in the best possible light (see figure 20).

Use this approach when setting and marking class tests. Arrange the test questions and marking system so that the marks range from 55–85 per cent instead of from the usual perhaps 15–45 per cent. These grade the students just as effectively and may give the class the encouragement it needs.

It is an excellent device to set a small test before a piece of instruction on a subject they all know *something* about, and then to set the same test after the instruction. This invariably shows definite progress, and creates a sense of achievement in the class. The test paper on p. 225 was set at the beginning and at the end of a course for instructors which lasted a fortnight. The marks were compared, and in each case there was a marked improvement. This test paper was more an object-lesson than a serious test, but it served to give the class a sense of achievement and progress on its second attempt. A progress chart for the whole class, although it entails a little work, is a useful motivator.

A pictorial representation of the various obstacles to be surmounted, for example tests and jobs to be done, has more appeal than a dry-looking syllabus. A picture of general class progress is not very effective in motivating individuals; for this, a system of individual progress reports is necessary. This topic is dealt with in chapter 6.

It has been found that people always work better if they know the results of their efforts, even if those results are below their expectations. This point is rather strikingly borne out in the two graphs plotted in figure 8, of two groups of students who were set to do mental multiplication. Group I was kept constantly informed of their results up to the first ten attempts, and were encouraged each time to beat their previous score. The men in Group II were not told their results at all for the first ten attempts. After each group had had ten attempts, the conditions were reversed. Group II was informed, and Group I was not, with the results shown in the graph (figure 8).

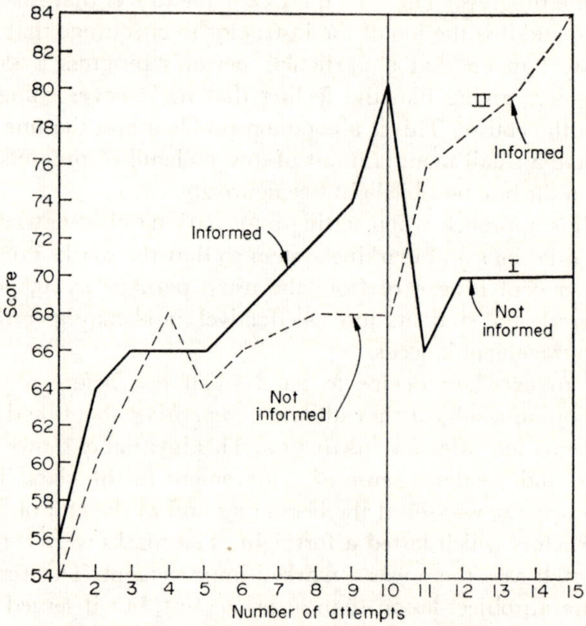

FIGURE 8 Progress related to informing of test results

This has been admirably summarised by Sir Frederick Bartlett: 'The common belief that "practice makes perfect" is not true, it is practice the results of which are known, which makes perfect.' Teaching machines exploit this principle, which is sometimes expressed in the form: 'Effective learning is quick "feedback" learning (see p. 110).

4. *Praise and rewards are better than blame and punishments.* An effective, but rather heartless way of teaching an animal to perform

tricks is to deprive it of a meal, and then use its desire for food as a motivator. Use a succulent morsel of food to entice it to do what you want, for example sit up or jump, and then instantly reward it with the piece of food. Here you have a motivated animal and an effective system of rewards for good effort. Learning is rapid in these circumstances—the animal soon learns to perform the trick, even when not hungry, provided there is some immediate reward after each performance. The reward should follow immediately in most teaching situations even if it is in the form of 'good that's the right answer'. This principle is basic to programmed learning.

There are two recognised ways of getting a donkey to move: the *stick* or the *carrot*. As we have already seen in chapter 3, you can induce men to travel along the path of learning far more satisfactorily by using the carrot.

Praise is more effective than blame, as are rewards compared with punishments. The reason is fairly simple. Praise and rewards are pleasant motivators, whereas blame and punishment have unpleasant associations and may make the victim dislike the instruction and the instructor.

Behaviour which is rewarded is positively reinforced, and is therefore likely to recur. The best planned learning situation provides for a steady cumulative sequence of successful behaviours, so that rewards should be reasonably frequent, immediate, and given for specific good responses whenever appropriate. Punishment is intended as a deterrent to wrong behaviour, and produces negative reinforcement by inhibiting undesirable responses, but it may not eliminate them entirely. It has a number of undesirable side effects and may produce anxiety and hostility. Negative reinforcement should therefore be avoided. Most people want to please those they like and respect, hence it is clear that praise is only really effective if it comes from an instructor who is liked and respected by the class.

These are merely general principles, and there are exceptions. You must know your man, as men are not alike and have to be treated as individuals. It may be right in some cases to give someone who is slack or undisciplined a dressing-down or a suitable punishment, making it clear by your manner that you have his interest at heart. A jolt may be what he needs, but be quite satisfied that you have tried, to the full, more friendly and persuasive methods without avail. The schoolmaster's reservation about using the stick is: 'When you punish a boy, have the boy's *end* in view.' Never be vindictive, and remember the principle firmly established by Gilbert and Sullivan of making the punishment fit the crime. Your attitude at all times must be *fair*, *firm* and *friendly*.

5. *Curiosity, realism, surprise and variety stimulate interest and therefore motivate the class.* One of the most effective ways of stimulating a child's interest and commanding his attention is to hold something behind your back and say: 'Guess what I have.' Use this kind of appeal to natural curiosity when appropriate in instruction. Keep the class guessing part of the time, for, although curiosity killed the cat, it help to keep a class alive. A puzzle is always a challenge, and a class should enjoy using its brains to solve a mystery.

This curiosity that arouses us and challenges us is at the bottom of our interest in the 'Whodunit' detective story, the fascination of conjuring tricks, quizzes, and 'Twenty Questions'. It should also play an important part in school science teaching.

The seasoned instructor, who holds up a piece of mechanism and then rattles off a word-perfect monologue in description of it, can come very near boring his class. A better approach would be: 'Here is a Snodgrass Pump. I want Smith to come out and turn the driving-wheel and the rest to watch carefully to see what happens. We shall puzzle it out step by step. Carry on, Smith. (Pause for some minutes.) Do you all see the cam in the corner? What is it doing? Davis?' The topic of curiosity is bound up with the art of asking questions, which is dealt with more fully in chapter 5.

We all enjoy the satisfaction of solving a problem. Intellectual achievement gives pleasure, and a difficult subject, if properly approached and handled, can become interesting for its own sake. A tricky point in electro-technics to a properly motivated class becomes like a mountain to a mountaineer—a challenge—and provides an equivalent satisfaction when conquered.

Realism and a sprinkling of surprises can be introduced into most pieces of instruction. Wherever possible, use the real thing to demonstrate instead of a drawing. Make your questions real and practical. Instead of: 'What is the formula for the minimum size of the supporting column in this structure?' try this: 'Measure up this material and let me know if you can make a supporting column from it for use on this job.'

Instead of: 'What are the rules of procedure for fault-finding on this set of machines?' try this: 'The machines are all out of order this morning. I want them in working order in fifteen minutes. Do the fault-finding systematically.'

If you are instructing in how to deal with any kind of emergency, make it as real as possible. Plan it and get an accomplice or two to help. If you stage an accident, get someone to be the victim and add a touch of realism.

6. *Keep the class as active as possible.* A class undergoing technical instruction should be given every chance of *doing*. Passive reception will not turn out trained technicians. Recall the stock joke of father on Christmas Day insisting on showing his frustrated and exasperated small son how to work the model train and the signals. Father is a real menace, because the small boy wants to *do* it himself.

Trainees want to express themselves and to take part in what is going on. Moreover, men often prefer their *own* way of going about learning a difficult topic. Within limits allow this latitude, but make sure that you are stimulating their natural curiosity and urge them to take part and do it themselves. This approach to an unexplored territory used by an intelligent and able instructor will prove of great value.

To a group of three or four men, who are ahead of the class, say: 'This is a new type of remote-control gear. I want you to inspect it, strip it if necessary, and find out all you can about it. I shall then ask one of you to explain how it works to the rest of the class.' Here we have possibilities for a little rivalry, curiosity and class activity, based on the 'discovery' method.

Give trainees productive work, but they should not work too soon with experienced employees on production as this may lead to discouragement over comparisons in output.

Teaching machines owe much of their success to the fact that they fully exploit the principles mentioned in the foregoing section: 1. learning is the task of the learner; 2. a strong element of motivating competition is introduced—'man versus machine'; 3. progress is graded, and there is an immediate assessment of the effort made; a man knows how he is doing; 4. a correct response produces a sense of achievement and is an immediate reward—teaching machine programmes of the linear type, which do not make at least 70 per cent correct responses a reasonable achievement, are found to cause dissatisfaction and fail in their purpose; 5. curiosity and interest are aroused by the series of graded questions or problems; 6. users of the machine are kept active and free from the dangers of passive reception.

Size of Classes. It has been said that the ideal size of a class is one, that is one teacher and one student, but this as we know from experience is almost unbearable—very boring after a short time.

The teacher is bound to dominate the scene and the student feels he is very much part of a teaching situation, with one person spending all his time personally concerned over what and how he is learning. A teaching machine is concerned just as much as you want it to be, and you can switch it off whenever you feel like it, and go it alone.

A class of one may be ideal in some respects, but would not contribute to attitude training and the character-forming objectives we have discussed. The obvious disadvantage of a large class is that there is little scope for individual activity and attention. Instructors soon discover that conditions of learning and the general atmosphere of a class varies considerably with the size of the class.

In a small group members rapidly know one another.

In a class of 20 or more there will be a number of sub-groups, and a number of individuals who can escape notice and will seek anonymity and enjoy seclusion at the back.

The methods of instruction for small groups include tutorials, discussions, syndicates, role acting, and these methods ensure participation in the lesson. In learning a skill, it has already been emphasised that about 65 per cent of the time should be spent in individual practice. This is impracticable with classes of more than about ten men. Experience has shown that an instructor teaching skills can give the necessary individual attention only to about ten men. Beyond this number instruction becomes more and more a one-way-traffic system and, when the class is fifty or a hundred strong, individual attention by the instructor becomes impossible. In general the smaller the class the greater the opportunities for individual attention, class activity, communication between teacher and taught, and the more successful the assimilation. The more advanced the instruction the smaller the class should be as each individual has more to contribute to discussions, and makes bigger demands on the instructor. Six to eight trainees seem to be a necessary minimum in order to establish a group structure which is so valuable in building class morale.

Twenty-five trainees per instructor is a maximum for the lesson type of instruction and eight to twelve is a maximum for demonstrations that are followed by practice. Unfortunately the size of the class is frequently determined by the size and number of rooms, and the practical facilities available.

For lectures, hortatory films and exhibitions the numbers are limited only by the capacity of the room or hall.

Films that are intended for discussion, or to be shown in small sections as a prelude to practical work, are better restricted to showing to a small class.

8. *Class comfort.* The physical conditions under which a class is working affect its morale and its degree of motivation. The two are closely associated.

A student is not well motivated if he is cold, or cannot see properly, or is squatting on an uncomfortable stool, or has the light in his eyes, or if there is an uncomfortable background of noise going on in the next room or workshop.

The more important points connected with class comfort are as follows:

The lighting should be such that all the class can see well with no hard shadows across the paper on which the students are writing. Inefficient lighting, particularly in the evening, can seriously hamper instruction. If the room has filament bulb lighting, try to seat the class so that the light is to their left or right rather than in front or behind. The best form of lighting is fluorescent lighting of adequate power and spacing.

Instructors in machine shops should be aware of the danger of stroboscopic effects with fluorescent lighting which can, for example, give a misleading impression of the motion of a machine such as a lathe.

Temperature. This should ideally be around 18°C. Make sure that none of the trainees is sitting in a bad draught, or has too much of the heater to himself. Ventilation is important, as stale air can act in conjunction with a boring instructor to produce a state of somnolence.

Seating. The class should have plenty of room and should have comfortable seats with a back-rest, and a table for a forward rest, and for writing.

Some technical training may have to be carried out in the open:

when this is the case, face the sun yourself rather than have your class suffer this discomfort. Sit the class down whenever convenient or plan the outdoor work so that unnecessary walking is eliminated. If the weather is bad, would the instruction be best indoors after all?

These are small points that affect the comfort of the class, but the good instructor will be on the look-out for them, as it is essential for the class to learn under the best conditions. Exercises to test or to develop the trainees' toughness or powers to stick at a hard job can be devised separately, and are important in some courses of training; do not introduce these special activities into lessons which are designed for acquiring knowledge or skills.

9. *Class morale.* Class morale is the spirit and atmosphere of the class, and determines whether the class is co-operative, motivated, cheerful, with a sense of class loyalty, or a group of unco-operative 'couldn't-care-less' grousers.

Class morale depends on a great many factors such as the attitude of top management, staff relations at all levels, the local youth environment, and the presence in the classroom of a few influential personalities—not the least of whom is the instructor, as we have seen in chapter 3.

The instructor–trainee relationship in a learning situation is sacrosanct, and a class under instruction should never be interrupted or disturbed by people popping in and out with notes or messages except in cases of emergency. Prearranged visits by senior officials, however, can help to give the class a feeling that they are an important part of the firm or institution.

The instructor should have full control of his class, and his authority should be recognised and upheld by all. A code of courtesy quickly establishes itself when its importance and value is appreciated.

If a good atmosphere arising from sound management prevails in the establishment or works, this will percolate into the training class, but your example and your attitude to the firm or training establishment, to your colleagues, to your class, and to each individual in it, will profoundly affect class morale.

A few minutes spent in informal conversation or discussion with a student in private, treating him as a responsible individual, may have a much greater motivating effect than several hours of impersonal contact and exhortation in the classroom.

The overall object in training is to fit a man for his place in industry and to make him better and happier at his job. The basic needs of the ordinary man in a works have been summarised by Sir Roy Pinsent, Bt., in his pamphlet *Productivity in Industry* as follows:

to make a secure living;
to be treated as a man and respected as a person;
to have a voice in matters affecting him;
to know that he is wanted and belongs;
to know that the boss and the 'top' really care what happens to him;
to have a chance to develop his highest skills and capacities;
to be assured of justice, openness, and absolute honesty at the hands of those who supervise and manage.

These needs have to be appreciated at the training stage, as they affect the overall object of training, which cannot be achieved without sound class morale.

Job enrichment

One of the many new training terms now in use is 'job enrichment' which improves efficiency on the job, and the satisfaction of the worker, by providing greater scope for personal achievement and advancement, more challenging and responsible work.

Ways of promoting job enrichment include: encouraging trainees to contribute ideas; involving them in the solution of problems; giving them a sense of responsibility; and providing a variety of jobs which require thought and decision.

A sound basis for the growth of job enrichment is a corresponding 'training enrichment' involving these methods at the training level; such methods are often known as participative training, see p. 158, chapter 8.

Sensitivity training, T-group training, group dynamics

These terms are now in general use to describe a number of participative learning methods whose aim is to improve trainees' skills in working with other people (see p. 140).

5

Assimilation

In chapter 3 an analogy was made between an instructor preparing a lesson and a chef preparing a banquet. The analogy can be stretched a little further, because instruction like food, is wasted unless taken in, digested, absorbed and used. Over-eating produces

Food:	A hungry man	eats	digests	absorbs	stores products in system ready for use.
Instruction:	A motivated man	attends	assimilates, or learns	remembers	stores in mind ready for use.

bad digestion. Cramming a subject of study produces bad assimilation and generally bad examination results. Instruction to be of any value must be assimilated and remembered. There are several factors which affect the ease with which we assimilate and remember.

Meaning and understanding

The instruction must have meaning and must be understood. A simple experiment will illustrate this. At the beginning of a period of instruction write on the blackboard a sentence of, say, six syllables :

'The boy ran across the road';

then tell the class you want to try a little experiment, but do not disclose what it is. Invite the class to read the sentence three times, and then rub it off. Immediately write six nonsense syllables, for example, yac deb ron sur zad pef; ask the class to read them three times and make no other comment; rub the syllables off and carry on normal instruction.

At the end of the period, ask the class to write down first the sentence and then the nonsense syllables. All will be able to recall the sentence—few if any the nonsense syllables, although equal attention was given to each. We remember best what has meaning and is understood.

Interest and attention

These have been discussed in chapter 6. We cannot satisfactorily remember anything unless we attend to it, and the deeper our interest, the more concentrated and effective is our attention. There is no very clear evidence for a faculty of memory, there are only *persons remembering.*

Depth of impression

Memory is closely correlated with the vividness or depth of the impression made by the instruction. Memory in this respect is analogous to the impression made at a recording of a gramophone disc; a faint impression gives poor reproduction and is easily obliterated. Hence the importance of making instruction as impressive and as vivid as possible, using the elements of surprise, curiosity, realism, class activity and competition as discussed in chapter 4.

Association of ideas

Memory is strongly aided by association of ideas. It is often fairly easy, after an interesting and rambling chat round a fireside or bar with some friends, to stop the conversation and rapidly trace back to its source the trend of the talking. This is made possible because one thought led to another, and the various topics were connected by an association of ideas. Ideas appear to be linked together in the mind, and whole groups of ideas are quite often recalled in association together. For the instructor the significance of this is that new ideas should be linked with previous knowledge and presented in a sequence that leads naturally from one idea to another. An important point can at times even be associated with a funny story, as a story is not readily forgotten and helps to recall the point by association.

The boy scouts have a test of observation and memory, known as

Kim's Game. Twenty-five assorted articles selected at random are placed on a table, the boy is allowed one minute to look at them, and must then write down all the articles he can remember. This is more difficult than it sounds, especially for those who do not have a good 'visual memory'. Some will find it helpful to make up a simple story bringing in each object to give some sort of meaning to the assortment. A man lights his *pipe* with a *match*, picks up his *purse* containing a *half-crown*, opens his door with a *key*, trips over his *boot-lace* going down the step, takes a bus and buys a *ticket*, buys a *carnation* for his button-hole, fixes it with a *pin*, looks at himself in a *small mirror*, *combs* his hair, then meets his girl friend, who has a *green necklace*, and a small dog with a *collar*, and so on, until all the articles have been included. Such a story even though fantastic, has some sort of meaning, and forms associations that aid the memory.

Frequency of repetition

There are two ways of getting a nail into a hard block of wood—either one hard, sharp, impressive blow, or a large number of small taps. Often topics of learning require a number of repetitions before assimilation and memory can be established, but, as mentioned on p. 60, mere repetition will not cause learning. Improvement in a skill comes only when the results of repetition give pleasure or satisfaction and confirms the skill acquired. When the results of a repetition annoy, then improvement comes by rejecting the unsatisfactory procedure.

Time

This fortunately makes us forget; fortunately, because life would be quite unbearable if we were doomed to remember everything we see, think, do or say. We should be truly thankful for our ability to forget. Good memory is often aided by selectively forgetting unimportant things.

Frequency of recall

Memory is helped by frequent opportunity for recall. I learn a telephone number today, but, if I do not have to use it for a week, I shall certainly forget it (unless it happens to be 1066, in which case it becomes associated with the only 'memorable' date in British History). If, however, I have to use it every day, I shall be able to recall it for several weeks or months.

Here is a summary of the factors affecting memory.

Factor	*How it affects memory*	*Notes for instructor*
meaning and understanding	memory very slight without these	make meaning clear, use simple words
interest and attention	memory is proportional to interest and attention	motivate the class inattention the instructor's sin
depth of impression	memory proportional to impression	be impressive, use surprise, curiosity, learning aids chapter 4
association	memory aided by association of ideas	good sequence, logical development humour, surprise
repetition	a piece of learning often has to be 'hammered' home	especially applicable to disconnected facts (sweat and test) to confirm
time	memory fades with time	regular and systematic revision, and tests with quick feedback
frequency of recall	recall strengthens memory	class should 'give back' from time to time use tests and assessment

While on the topic of memory, it is worth noting that facts and figures are in themselves difficult to remember and, unless of vital importance to a man on his job, should be relegated to the *could know* part of the instruction. The instructor should encourage trainees to look things up, and tell them where they can find information. Dimensions of parts of machines and the size, pitch and core diameter of various screw-threads are more easily looked up than remembered. Time is more wisely spent in making sure that general principles are known.

There are of course in all branches of technology many facts and figures which must be carried in the head and the learning of such facts is best done by a process known as 'sweat and test' : the 'sweat' indicates that the man is well motivated and makes real effort; the 'test' part confirms the learning and provides opportunity for recall, giving back followed by more repetition. These together ensure assimilation.

Repetition, variation and insight

We have all experienced the sudden flash of insight that comes when we ponder over and grapple with a particularly difficult problem. Instructors are generally familiar with the change on a trainee's face from a look of bewilderment to one of triumph when in a flash of inspiration the 'penny drops', as the process is sometimes colloquially called. Whenever there is difficulty in getting the class to see a complicated point, insight is assisted by repeating the reasoning with some variations or by a new approach.

Class activity assists assimilation. Points that have been worked out by a class make a deeper mental impression and are more readily assimilated and remembered than points that have been spoonfed to the class. The simplest, and perhaps the most effective, way of making a class or individuals work things out for themselves is by means of *questions and answers.*

Questions and answers

Questions are asked by the instructor to make the class think or to test knowledge and assimilation. They produce immediate feedback and provide a means of making a continuous assessment of the progress of a class or an individual.

They are an essential part of the teaching-learning process. In fact good technical instruction is mainly a matter of asking the right questions in the right sequence. Programmed learning is based on this principle.

A class of prospective instructors on a training course, might profitably have a discussion on the use of questions in teaching and learning. This would help to systematise the knowledge of the class, and would be an appropriate demonstration of class participation.

Points for discussion would include :

Why ask questions of a class?

What are the main kinds of questions used in teaching and training?

How do questions help the trainee to learn?

How do questions help the instructor to teach?

When should questions be asked?

The results of such a discussion might be written up as a chalkboard summary on the following pattern : (Questions by the class are dealt with on p. 77.)

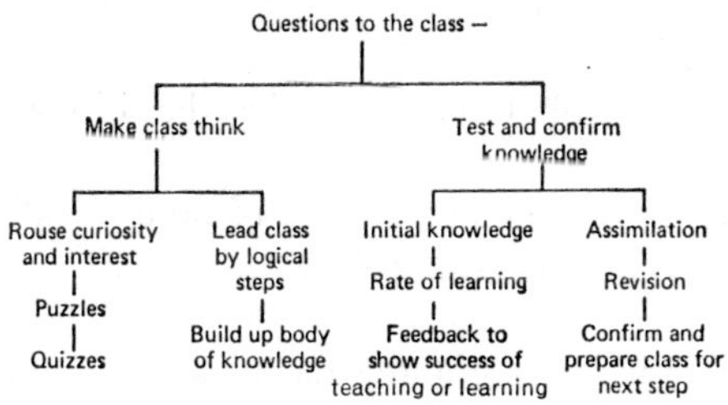

FIGURE 9

The following are some examples of questions.

To arouse interest, we have, for example :

What is the thrust of each of the four engines of the Concorde aircraft? Discuss this and calculate the acceleration during take off given the mass of the aircraft, using SI units.

To build up step by step some principle, question-and-answer might proceed thus :

Q. In this circuit (for the megger), suppose I made x, the unknown resistance, zero. How will this affect the current in the coil B? . . . pause . . . *Mr. Smith.*

A. It will be large compared with that in coil A.

Q. What will happen to the pointer on the instrument? . . . *Mr. Jones.*

A. It will swing to the right.

Q. Now suppose I make x several million ohms. How will this affect the current in coil B? . . . *Mr. Simpson.*

A. The current in B will be small compared with the current in A.

Q. What will happen to the pointer now? . . . *Mr. Baker.*

A. It will swing to the left.

Q. How can we use these ideas to make up an instrument to measure very high resistances? . . . *Mr. Green.*

Thus the class takes part in working it out for themselves, step by step, in a logical order.

Here is a 'wrong way' demonstration playlet on questions and answers. (More can be made up for instructor training courses.)

Subject—Elementary Marine Engineering

I. What must ships have to make them go? . . . *Rogers.*

A. A captain.

I. Certainly. What do you say, *Snelgrove*?

A. An engineer.

I. Quite true, but what actually makes the ship go? . . . *Taylor.*

A. Engines.

I. Yes, quite so, but the engines are of no use without something else. What is the something else? . . . *Edwards.*

A. Coal or oil.

I. Correct, but it was not quite what I wanted. Let me try another approach. What is it which goes round and round at the stern of a ship? . . . *Roberts.*

Roberts having lost interest rises to the occasion and replies :

A. The log-line.

I. You are all very stupid. They are beautifully curved things beginning with S.

A. Wag at the back of class : *Sharks!*

I. No! I'll have to tell *you.* Screws, commonly known as propellers, and we are going to study ships' screws this morning.

There was no trace of reasoning or class activity used in producing these answers. The instructor was merely playing a very dull guessing-game, and got what he deserved. The class became slightly bored and ended by pulling his leg—a very poor introduction of the new topic.

Questions to test knowledge, rate of learning, or state of assimilation are generally ordinary straightforward questions, for example :

What tests must be carried out on completion of any electrical installation?
What is the current-carrying capacity of a 19/·064 cable!

Some questions combine a test of knowledge with a stimulus to interest and thought, for example:

How would you adjust the cam which operates the fuel pump in order to make this engine run in a reverse direction?

There are several points to observe in putting questions to a class.

1. *Name last.* Put the question first to the class, letting your eye rove over the whole class. Pause for a few seconds, then name the person you wish to answer it. This encourages everyone to listen and think out an answer in anticipation of being asked.

2. *Spread.* Invitations to answer questions should be spread evenly throughout the class. They should not be restricted to a few bright members of the class or to the dull elements. Let there be no obvious system, such as going round the class, as this defeats the object of 1. It is sometimes helpful to tick names on the seating plan (chapter 3, p. 31) to ensure that questions are spread evenly and fairly.

3. *Phrase carefully.* The meaning must be quite clear, and expressed in simple language; not 'What are the consequences of forcefully elevating the small spherical projection?' but 'What happens when the small knob is pushed up?'

4. *Wrong answers.* Wrong answers are valuable in that they very often indicate weaknesses in your instruction. Use wrong answers to clear up hazy points and in planning revision.

5. *Steps.* Step-by-step questions should be such that there are no big jumps from one piece of reasoning to the next.

6. *Answer must be heard.* It is a good plan to repeat answers so as to make sure that the whole class has heard.

7. *Avoid questions that have an obvious answer.* They waste time and stimulate no thought, for example:

'What is a starter used for?'

Echo questions come in this class. These are questions that have as an answer, an echo of your previous remark, for example: 'When I push the lever forward, the pin engages the low gear wheel. What happens when I push the lever forward? . . . *Thomas*.' This is sometimes more often used to find out whether or not a man is asleep. It is a dodge that may have its uses in a nursery class in providing class activity at a low level but it should not be allowed out of the nursery class.

8. *Avoid questions which have several equally good answers*, and the class can give the answer you want only by guessing, for example: 'What should all lorry drivers carry with them?' The answer might be licence, mate, spare wheel, petrol, works order.

9. *Avoid questions that call for Yes or No as an answer*. These are of doubtful value unless followed by more searching questions, because a trainee has an even chance of guessing the right answer.

The same applies to questions that have only two possible answers, for example: 'Does the magnet increase the current or decrease it . . . *Williams*?'

10. *Avoid trick questions*, for example: 'How many cylinders has a four-stroke engine . . . *Bell*?' Riddles and catches, unless really amusing, irritate the class and achieve nothing.

11. *Avoid questions to which the class cannot be expected to give an answer*, for example a formula the class has not met before.

12. *Avoid questions which are a test of powers of expression*, for example 'How do you wind an armature . . . *Fairfield*?', or, even worse, but in the same category, 'How do you keep your balance on a bicycle . . . *Hayter*?'—unless you *wish* to test his powers of expression.

You cannot test a man's *skill* by means of questions; you test a man's knowledge of a practical job far better by getting him to *do* it, than by asking him to tell you how he would do it.

Questions are suitable only for testing knowledge. It may occasionally be necessary to ask a man how he does something. If so, let him explain in his own way and, even if you get a very muddled answer, do not assume that the man does not know how to do it.

13. *Avoid general questions* that are not definite and *to the point*, for example 'Do you see what I mean?' It is better to use test questions at intervals to make sure that the class is with you.

Do not ask questions for the sake of asking them.

Reference has already been made on p. 38 to the bad habit some instructors develop of enquiring without much purpose, 'Right?' every few seconds.

14. *As a general rule, avoid putting questions to a man who is busy practising or demonstrating.* If you want to make a point put a question to the rest of the class, or stop the man on the job before you ask him. He can then give you his undivided attention.

15. *Give due credit for good answers*, even though they were not exactly what you wanted. A word of praise to the less-bright trainees does much to build confidence.

Questions by the class

The class should be given opportunities to ask *you* questions. It is important not to discourage the class from asking questions by such damping comments as 'I can't understand why people ask that silly question,' or 'You have clearly not been paying attention or you would know the answer to that one.' Such an attitude will put off many good questioners.

Questions by the class which anticipate later work could be written down in the instructor's notebook and taken up at the appropriate time.

Welcome and encourage questions. Frequently the class will remind you of a good point by a question. Thank them for it, for example 'Collins has raised a good point, he asks, etc . . .' 'Can anyone give the answer? Can you . . . *Steele*?' When you receive the correct answer from the class confirm it as correct and repeat it so that you are certain all have heard it. If you do not know the answer yourself, say so, and promise to find out. But remember to find out.

Trainees in a learning situation will be bristling with questions both in class and during practical work, many of which they can be guided to answer for themselves.

The instructor's job is to help the process. He must therefore understand how people learn, and what makes them want to learn. How, when, why, where, what and who have to be anticipated and thought out by the instructor who should then help the trainees to find the answers themselves as far as possible.

The instructor must be able to communicate to the trainee *what* he has to do, *why* he has to do it, *how* he has to do it, remembering that telling is not teaching and listening is not learning. Telling must be combined with showing, explaining, practising, discovering and questioning, before a skill can be thoroughly learnt.

Questions beginning with *what, why, when, how, where* and *who* form the basis of communication in a learning situation. Without them there can be no feedback.

These six words were called by Rudyard Kipling his 'six honest serving men'. They serve the cause of teaching and learning in this technological age very specially.

> 'I keep six honest serving men,
> (They taught me all I knew)
> Their names are What, and Why, and When,
> And How and Where and Who.'

Skills

So far we have dealt only with the assimilation of *knowledge*, but a great deal of technical training is concerned with imparting skills.

A practical skill is acquired, as we mentioned on p. 23, through a kind of muscular memory using our kinaesthetic sense. After we have successfully carried out a series of intricate operations, our muscles seem to remember how to do it and, even after a lapse of some weeks, we find we still have a fair amount of that skill. This skill–memory is much more durable than our memory for facts. Indeed, once we have learnt how to ride a bicycle, or to swim, or to ride a horse, we never forget. The skill is not actually stored in our muscles but becomes a pattern in our mind and nervous system. The brain is the coordinator and controls all our activities.

Signing your name, for example, might be regarded as the result of a well-trained set of wrist and finger muscles but, if you attempt to sign your name in the air with your big toe, you will find yourself going over exactly the same form that your pen makes. The skill is part of a mental pattern. Although this skill–memory is more retentive than ordinary mind–memory, it nevertheless does decrease with time, but a few trials will soon restore the former standard.

Here are a few simple points about skill–memory. Learn the skill correctly the first time. If a faulty or awkward movement is acquired at an early stage, it becomes very difficult to get rid of the fault later. Cricketers, boxers, golfers, field and track athletes know this very well.

Expert demonstration is essential right from the start, and the objective should be accuracy and correctness first and then speed. If, however, the demonstration is done too slowly, the rhythm and form of the desired skilled movement is sacrificed.

An over-emphasis on accuracy can lead to a set of deliberate movements that prevent the ultimate achievement of desirable speed. Training must be related to productivity. Most productive workers have to aim at quality, but quality in quantity—time is important. You should therefore urge students to concentrate on accuracy of working first, but ensure that the correct movements are being practised, and that no time is wasted. The best way to do this is to make a skill analysis. This involves a careful study of exactly what each hand does, in what sequence and which fingers are involved in which position. It further involves defining the key points to be looked for. The whole study is based on how an experienced and efficient worker does the job. A skill analysis should be carried out in association with a job analysis and generally requires specialist treatment, but any instructor before starting to teach a skill will be greatly helped by making even a simple analysis of his skill. A fuller treatment of this topic is beyond the scope of this book.*

As in mind–memory, skills need time to be assimilated. Our tennis or skating will not be improved if we play *intensively* for months on end; staleness sets in and we reach a stage when there is no improvement. Several weeks or even months of rest are often the best thing for some athletic skills. It has even been said that we learn to skate in summer and swim in winter. This is simply a paradoxical way of saying that the muscular skills we acquire in summer may be assimilated thoroughly in the winter ready for use the next summer.

Literacy and skills

Educationists concerned with technical education in the developing countries have noticed a marked degree of correlation between literacy and ability to learn psychomotor skills. This correlation would seem to be reasonable on the grounds that: 1. a literate person is more experienced in methods of communication; 2. a literate person has ecquired literacy through the exercise of a wide range of cognitive and psychomotor skills such as are involved in reading and writing; and 3. literate persons are already experienced in learning how to learn.

* For further study consult standard works on this subject such as *Industrial Skills* by W. D. Seymour.

If there is such a correlation and transfer, it would lend support to the view that a good basis for eight years of general education is of fundamental value and importance in all forms of skill training at all levels.

Memory aids

Learning based on fundamental principles and logical reasoning which is understood, is far more valuable and easily recalled than learning acquired by rule-of-thumb methods, including memory rhymes or mnemonics, which can so easily be forgotten or distorted. Rhymes and mnemonics are, however, useful when no logical principle is concerned, for example 'Thirty days hath September, etc.' The colours of the spectrum are contained in VIBGYOR. Sailors use memory tricks to remember their navigation; the topmarks of middleground buoys are remembered by 'to sea for diamonds and home for Bass'. An example of a pleasant association of ideas.

As a general rule, discourage learning like a parrot, and certainly avoid teaching like one. Rote learning can be disastrous,

'For thought they wrote it all by rote,
They did not write it right.'

Give every opportunity you can to trainees to put into practice, and to express, what they have learned. Expression deepens the impression.

Schoolboy 'howlers' are the result of bad assimilation. A boy, after several lessons on general science and geography, was asked in a test: 'What causes the tides?' The reply given was: 'The tides are caused by the rays of the moon striking the surface of the sea at an angle of $23\frac{1}{2}°$ Fahrenheit!' A good example of bad assimilation.

The steps generally followed in the learning process are: desire to learn—motivation; interest; attention; participation; assimilation —memory; application; assessment of achievement; feedback.

Ending a period

The ending of a period of instruction is a matter of some importance, and gives the instructor an opportunity to give any finishing touches to the processes of assimilation and to test how far the object of the instruction has been achieved. The instructor, when about five minutes from the end of the period, might ask, 'Any questions?' Then ask the class some. Finally, give a summary which should be a brief account of the main points of the *'must know'* part of the instruction.

If the period has been devoted to learning a skill, then the ending might be along the following lines :

'Any questions?'; final practice of essential skills; instructor comments on progress made in relation to what has to come next. Some notes on the closing of a practical period from the point of view of workshop or practical class management have already been given at the end of chapter 3.

The subject of assimilation which results from sound learning cannot adequately be dealt with in one chapter, as it enters into every aspect of instruction. Testing and progress-reporting play a part in assimilation, and they are considered in the next chapter.

6

Assessment

Assessment is an integral part of the teaching–learning process. The objectives of a training programme are achieved by good motivation, sound teaching, activity learning, continuous assessment and feedback, operating in a unified process (see the training function diagram p. 111). Tests, examinations and progress reports play an important role in any course of instruction; they can be used to help the class and to provide valuable information for the instructor.

For the class, a test can : (*a*) provide a goal to be reached, and promote a sense of achievement; (*b*) confirm what has been learnt and so aid assimilation by reinforcement; (*c*) motivate the class by competition, as each trainee wants to do well and excel; and (*d*) enable each trainee to know how he is faring on the course, what he is good at and what his particular weaknesses are.

For the instructor, a test can : (*a*) make clear what the class knows at the start, if set at the beginning of the course; (*b*) grade the class by ability; (*c*) measure progress and how much of the teaching has been assimilated, and so reveal weaknesses in the instruction—well-maintained records allow training officers to compare different training techniques and so to experiment with successful methods; (*d*) show how much revision is necessary and whether the object of the instruction or the course has been achieved for individuals as well as the class as a whole; and *e*. some firms have found progress reports valuable when considering a trainee's future specific job. The main purposes of tests and examination are shown in the chalkboard summary in figure 17 on p. 110.

The instructor should not use tests merely as a means of stimulating a lazy class, and, although such tests may occasionally be necessary, it is more valuable to design tests to give confidence and encouragement. This does not mean that tests have to be easy. Preliminary tests should in fact be a little harder and more exacting than the final test, for much the same reason that an athlete training for putting the weight will practise from time to time with an over-weight shot.

A good system of testing and reporting is one which gives to the serious student a sense of progress and achievement. This can be done by testing at the beginning, at intervals throughout the course, and at the end of the course.

As a general rule tests should be set a few days or a week or two after the period of instruction. If you test a class immediately after a piece of work, there is a tendency for assimilation to be arrested because a student may rely on a short-term memory and forget everything the test covered as soon as it is over. As we have seen earlier, good assimilation facilitates good retention. After a test has been completed the instructor should *not* encourage the attitude that the test was the end of that part of the work; he should make it quite clear that each test is a means of consolidating a position gained, before an attack on the next part of the syllabus, which will lead up to the next test, and so on.

It is important to give trainees the result of a test and to hold a post-mortem on the result, within a day or two of the test. The longer the interval between test and result, the more remote the test becomes in the minds of the trainees and the good effects of timely praise or censure rapidly diminish. Post-mortems on tests are excellent for teaching and confirming what has been learnt. We saw in chapter 4, p. 59, that the whole attitude of a class towards its work can be influenced by the way it is tested and kept informed of results. This principle, as we have seen in chapter 4, p. 63, is exploited in the use of teaching machines or in programmed learning.

Testing skills and techniques

In testing a skill or a technique, make the test a practical one. This point can be made during a course by asking an instructor trainee, 'Mr. Jones can you ride a bicycle?'

Mr. Jones, 'Yes.'

Instructor, 'Would you please tell us or describe how you ride a bicycle?'

Mr. Jones is immediately in trouble. He will make a fool of himself by either trying to describe the skill, or going through the motions of riding a bicycle, but the instructor will have no idea whether or not Jones can ride a bicycle. A silly question gets a silly answer. Many examination questions are equally silly—or invalid. The question to test the practical skill should have been. 'Mr. Jones, would you please show us that you can ride this bicycle.' Have one ready for the purpose. You test a man's skill if you have him in the shop and say: 'Join these two cables.' You probably test his powers of expression if you set a written question or ask him orally: 'Describe how you would join two 19/.064 electric cables.'

A man may get good marks for his oral or written question, but make a poor showing if confronted with the job, and vice versa.

In general avoid tests that require the essay type of answer. This can be done by using one or more of the short objective tests on p. 99, such as the 'multiple choice' type of test which has several advantages. It eliminates writing and is therefore not concerned with testing a candidate's powers of expression. It takes only a few minutes to complete, and is rapidly and easily marked, but it requires care and skill to set. Results are independent of the marker, so the test is quite objective.

Appendix I gives an example of a test (not to be taken too seriously) which has been used by the author at the beginning and again at the end of a course for instructors to illustrate: 1. the use of a multiple-choice type test; 2. the use of a test to find out what the class knows at the start; and 3. the use of a test to produce a feeling of progress and to grade trainees.

A class of twenty-five can be tested in ten minutes, the papers can be marked quite objectively by means of a grid, and the results given out in about half an hour of the close of the test. This incidentally demonstrates the motivating effect of quick feedback.

Points to note in setting this type of test

There should be a large number of questions. About thirty is a satisfactory number in order to derive a fair score. Answer sheets can be provided, and candidates can indicate their selected answer with a ring thus:

1.	A.	2.	A.
	B.		(B).
	(C).		C.
	D.		D.

Each of the alternatives offered should be a reasonable and possible answer. Alternatives having obviously wrong or nonsensical answers should not be used. The position of the correct answer should be varied haphazardly. A choice of four answers to each question is a good number; a choice of two encourages guessing.

In order to correct for the possibility that some of the right answers may have been good guesses, the actual number of correct answers scored has to be slightly reduced, in order to obtain a score which will penalise the guesser, but be fair to the candidate who obeys the instructions and leaves questions to which he does not know the answer. If a man gets n answers right and m answers wrong in the test given, it is reasonable to infer that out of the n he got right, that $m/3$ of these are probably attributable to lucky guesses.

How can one determine that a correct answer was a guess? If each answer can be answered in four ways, then the probability of a man guessing the right answer is 1 in 4, and of getting the wrong answer is 3 in 4. Thus, with a large number of questions, for every four questions guessed at we should expect three guesses to be unlucky and one lucky.

The final score is calculated as follows:

$$\text{score} = \text{number right} - \left[\frac{\text{number wrong}}{\text{number of choices} - 1}\right]$$

Thus in a test of fifty questions, if a candidate had

total correct	15
total incorrect	31
not attempted	4 (ignore these)

we should take the fifteen wrong answers as being a measure of the guesswork indulged in, and argue that, as he got fifteen wrong by unlucky guessing, he probably got one-third of this number, i.e. five, right by lucky guessing, and he should be penalised for this by five marks. His score would therefore be $31 - 15/3 = 26$. Had the candidate not guessed at all but left the fifteen questions he got wrong, his score would have been thirty-one.

Suppose fifty questions are set with only two choices offered in each question, and suppose a candidate scores twenty-five right and twenty-five wrong, his score by the formula above would be $25(-25/2-1) = 0$. This again seems fair because a candidate knowing nothing at all could score twenty-five right by guessing. If he had scored twenty-five right and got none wrong, his score would have been twenty-five.

Thus, if there are N questions, with p choices in each, a man gets n right and m wrong.

His corrected score is $n - \frac{m}{p-1}$ out of N. (1)

If a candidate knows nothing about the subject and guesses all the time, he is likely to get N/p right, and $N - N/p$ wrong. His score by substituting in formula (1) will be 0, which is what he deserves.

There are several other ways of avoiding a written essay type of test. The *one-word answer test* is one which merely calls for the missing word to be supplied, e.g.:

The usual limits of accuracy for mass-production grinding is . . . (0·001 mm).

For lathe work the cutting edge of the tool should be . . . (0·75 mm) below the centre of the work.

This test is only useful when testing lists of names, dates or sizes, etc. Care must be taken in setting it, otherwise long answers or explanations that are not called for will be given.

Practical tests

In order not to waste time, everyone under test must be employed. Tests should be interesting, and closely connected with the actual work the men are likely to have to do later.

In an important practical test it is generally desirable to test each trainee in four or five different kinds of job. This could mean that for a class of twenty men about eighty or more sets of testing gear would be required. The 'county fair' type of test economises in sets of apparatus; everyone is kept busy and all apparatus is in full use. Divide the men into, say, four batches and have four sets of jobs. Each batch has five minutes on a job after which each batch passes on to the next at a given signal.

Figure 10 is an example taken from the Army booklet *Good Instruction*. The work can be judged by going round and making rapid notes on how each man tackles his job. In addition, each man can be asked to write down on a card in three or four words the key to the practical problem set, e.g. *density of acid* 1·122 *at* 20°*C* or *distributor spring broken*. This assists marking but it should not become a test of expression.

Tests should be regarded as a part of training, or as a training aid, and not as a necessary interruption to be endured. A test is frequently of great importance in deciding a man's future and should be set by men of experience.

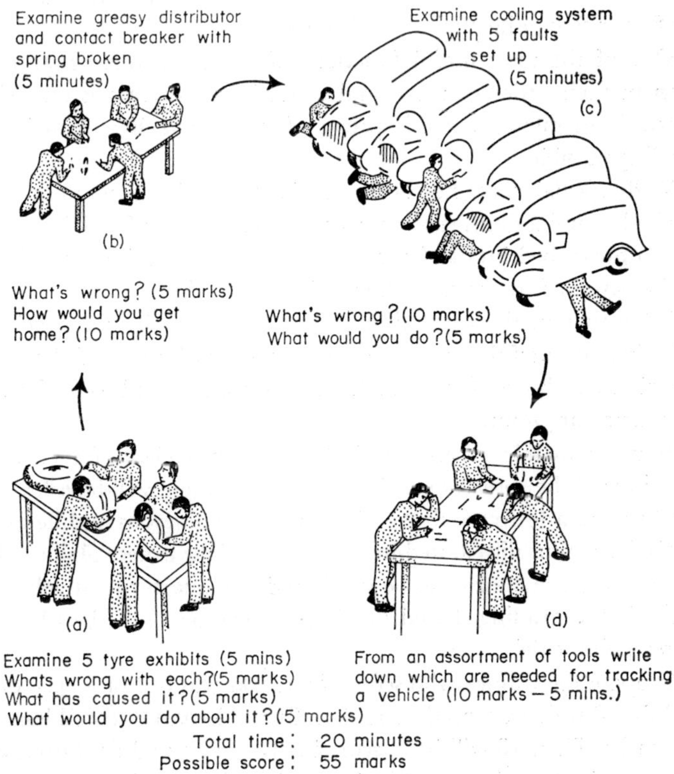

FIGURE 10 Example of a 'county fair' test

Unreliability of marks

Many instructors, and indeed schoolmasters and education authorities, tend to regard marks with a certain reverence as if they possessed a kind of magic, whereas they are in fact generally nothing more than a convenient means of giving a man a rank in his class. Rarely can marks give a correct idea of an absolute standard reached.

For example, a man who says that he scored 65 per cent in an examination conveys very little about his real worth in his class or the actual standard he has reached, unless we know a great deal more about the examination. The pass mark in this case might have been 90 per cent or 50 per cent; the examiner might have been harsh or benevolent; the student might have been bottom of his class with his 65 per cent or he might have been top!

One of the most useful functions of an examination is to grade men in their order of merit and a good examination is one which succeeds in giving a good spread of marks which will make this grading possible.

With special care, however, standards can be fixed in specially devised examinations, such as practical tests calling for skill in working to prescribed limits of accuracy, so that the marking is objective and a pass mark represents a fair and fixed standard.

In testing used as a training aid, it is generally more reliable, effective and useful to measure a man's ability by that of the rest of the class: to know, for example, that a man is ninth out of 100 in his particular group.

It is reasonable to assume that ability in a class of men under training will be spread over the class so that there will be a few very bright men, a few very dull and a far more numerous middle group. If ability is thus distributed in a class, we might expect about half to be average, one quarter to be good, and one quarter relatively poor. This simple grading, based on three categories, is said to be on a 'three-point scale'.

Suppose the trainee who scored 65 marks was one of a class of 32, and that 8 trainees scored more than 61 marks and 8 scored less than 43 marks. Then 16 will have scored between 43 and 61. Suppose 52 is the median mark, that it is the mark which divides the class into two halves, so that 16 get above 52 and 16 get below 52. Now we know a great deal about the examination and the trainee's mark of 65 tells us roughly where he stands in the class.

The marks 43, 52, 61 divide the class into 4 quarters called 'quartiles'. The trainee with 65 marks is placed in the upper quartile and he is in the top 25 per cent of the class. The range of marks between the upper and the lower quartile (in this case $61-43 = 18$) is a useful indication of the spread of marks in a test or examination, and is called the *inter-quartile range*. For practical purposes, the *median mark* is approximately the average mark and, if we *add* half the inter-quartile range to the average mark, we get the upper quartile and, if we subtract it, we get the lower quartile. Now suppose the trainee gained his 65 marks in a test in which the average mark was 72 and the inter-quartile range was 8 marks. He now finds himself not in the elite *top* 25 per cent of the class, but in the *bottom* 25 per cent! Beware of marks without knowing at least the average mark and the inter-quartile range!

To be told the mere mark of a trainee in an examination tells you nothing—unless you know the average mark (which tells you whether he is in the top half or bottom half of the class). It is better still if you are told the inter-quartile range of the marks, you can then place the trainee in one of four quarters of the class. In the table of marks on p. 92 the inter-quartile range is 10.

There can be considerable difference in the marks obtained from different instructors applying their own standards; but, if the students are ranked by ability within the group, a much more consistent picture is obtained. This is indicated by the following example:

Students	Instructor I (*Strict*)		Instructor II (*Easy with marks*)	
	Marks	*Rank*	*Marks*	*Rank*
Brown	8	9	10	9
Jones	6	8	9	8
Ingram	4	7	8	7
Hitch	4	7	8	7
Evans	3	6	6	6
Able	2	5	6	6
Dean	2	5	6	6
Cook	1	3	3	3
Field	0	1	2	2
Green	0	1	1	1

There are many ways of grading a class, making the assumption that the abilities of the students are distributed normally.

To the mathematician, the 'normal' is a symmetrical distribution in which values near the average value appear more frequently than any other, as shown graphically in figure 11.

Grading is a matter of dividing the area under the normal distribution curve into strips of equal width. The height of each strip represents the number of students in each division or grade. The number of strips is usually a small odd number: 3, 5, 7 or 9.

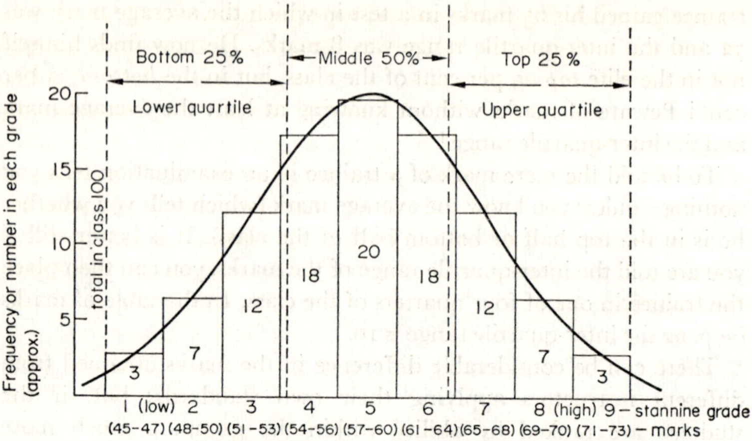

FIGURE 11 Distribution curve of examination marks scored as shown in the tables on p. 92

The following two paragraphs are for those who may be interested in this branch of statistics.

It will be noticed that the shape of the curve is bell shaped, but when the marks have a big range or spread, for example 0 to 100, the curve has a flatter hump. A small spread of marks say 60 to 75 gives a higher hump. Marks thus deviate on either side of the mean and a measure of the spread is called the deviation. We have used one way of measuring the spread—by means of the inter-quartile range which chops the class into a top and bottom 25 per cent with 50 per cent in the middle. A mathematical measure of the spread of a normal distribution is called the standard deviation. It can be regarded as the width of the curve.

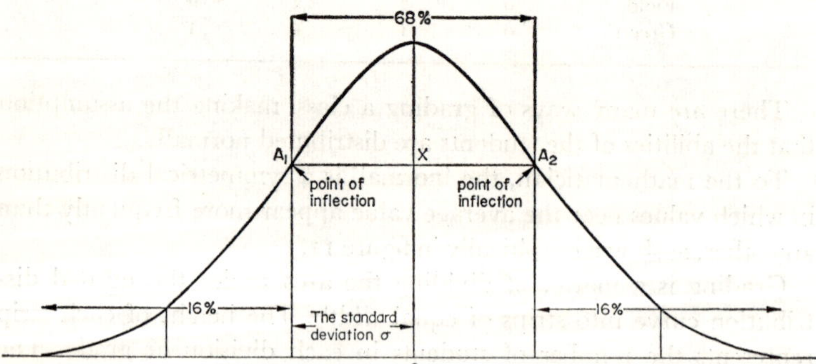

FIGURE 12 Normal distribution curve

It will be noted that the normal distribution curve (figure 12) has, at two points A_1 and A_2 on either side of the average, points of inflection which are mathematically convenient points for measuring the width of the curve. A_1X is called the standard deviation denoted by σ, The standard deviation $\pm\sigma$, on either side of the median, chops the class into a top and bottom 16 per cent with 68 per cent in the middle. In the diagram figure 11 the inter-quartile range is 10. The semi inter-quartile range is 5. The standard deviation is 7·5. Very roughly the standard deviation is about 1·5 times the semi inter-quartile range.

Grading

A simple grading on a three-point scale of good, average and poor was mentioned above. If the grades 'very good' and 'very weak' are introduced, this becomes a five-point scale in which the following might be expected: 5 per cent very good, 25 per cent good, 40 per cent average, 25 per cent weak and 5 per cent very weak. On a seven-point scale the following is a likely distribution: 5 per cent excellent, 10 per cent very good, 20 per cent good, 30 per cent average, 20 per cent fair, 10 per cent poor and 5 per cent very poor.

In dealing with large classes or groups doing the same course, a nine-point scale is useful (with the width of each strip $\sigma/2$). Thus nine strips cover a spread of $4\frac{1}{2}$ times the standard deviation and but, as the use of descriptive terms by different assessors varies considerably, grade numbers ranging down from nine to one become more convenient and are capable of definition for progress reporting. In a class or group of 100 students on the same course, the numbers that can be expected to fall into each grade are roughly as follows:

Top	3	...	Grade	9
Next	7	...	,,	8
,,	12	...	,,	7
,,	18	...	,,	6
,,	20	...	,,	5
,,	18	...	,,	4
,,	12	...	,,	3
,,	7	...	,,	2
Bottom	3	...	,,	1

The figures on the left have been calculated from the properties of the normal distribution curve shown in figure 11.

		(a) Score obtained	(b) Number getting this score (frequency)	(c) Cumulative total from bottom, giving percentile scores	(d) Total in grade	(e) Stannine grade
↑		73	1	100		
		72	2	99	3	9
		71	0			
Top 25%		70	3	97	8	8
		69	5	94		
		68	3	89		
		67	4	86	11	7
		66	2	82		
		65	2	80		
↓	Upper	64	5	78		
	quartile	63	3	73	17	6
		62	5	70		
		61	4	65		
		60	4	61		
Interquartile		59	8	57	21	5
Range = 10		58	5	49		
		57	4	44		
		56	7	40		
		55	5	33	18	4
	Lower	54	6	28		
↑	quartile	53	5	22		
		52	5	17	12	3
		51	2	12		
		50	3	10		
Bottom 25%		49	3	7	7	2
		48	1	4		
		47	1	3		
		46	0		3	1
		45	2	2		
↓	Total trainees		100		100	

Upper quartile = 63·5 Lower quartile = 53·5

Inter-quartile range = 10 marks

In these results the median is approximately the average mark = 58·5

Grading on a nine-point scale is known as stannine grading, the term stannine being derived from 'standard' and 'nine'.

As we considered a class of 100 students, the numbers falling in each grade can be readily calculated for larger or smaller classes. Note that about 50 per cent of the class will normally appear in grades 4, 5, 6, 10 per cent in the two lowest grades, and 10 per cent in the two top grades. This should serve as a rough guide for a class of any number.

To turn a 100 trainees' test score into stannine grades, write down the full range of the marks column (a) in the table on p. 92, and then write down the number of students getting each score. Now divide into groups as near as possible to the numbers 3, 7, 12, 18, 20, 18, 12, 7, 3 as shown in column (d).

The actual distribution of ability in the class may not correspond to a normal curve: a few *very* brilliant people among the top ones might make the representative curve not normal but 'skew'. In such cases, percentile grading would have the advantage of diminishing the skew effect of the unusually high marks earned by such people by still showing the *relative* grading of the group.

The column (c) shows the number in the class of 100 trainees, who obtained the score or *less* than the score shown in column (a). Thus the figures in column (c) give precisely each trainees' position in the class by means of a percentile score or a percentile rank. Note the top trainee gets a percentile rank 100, the bottom two trainees get a percentile rank 2. The lower quartile comes at approximately 53·5, the upper quartile at approximately 63·5. If a trainee's percentile score is say, 33 this means that he stands in rank 33 per cent up the list from the bottom. Percentiles are useful when dealing with large numbers for progress records, for selection purposes, for adjusting pass marks or for adjusting discrepancies between different examiners.

Grading in the ways outlined will usefully make a student conscious of his place in class which is an important measure of his ability. Thus, if his grades in successive tests are 4, 5, 5, 6, it is reasonable to infer that he is improving; if 8, 7, 7, 6, he is falling off. Raw marks may not reveal these trends.

It is true that grading does not show the progress of the class as a whole, but it can be supplemented by standarised tests from time to time to check that the class as a whole is up to the mark—and at least equal to previous good classes—using the same standard tests.

It is important in setting a test to select questions that will test the full powers of all but the top two or three, and yet be within the ability range of all but the very bottom two or three. This ensures a good distribution of scores. Examination mark lists often give more information about the skill of the examiner in his job than that of the examinee in his.

The following histograms of tests set to a class of twenty students will illustrate this point.

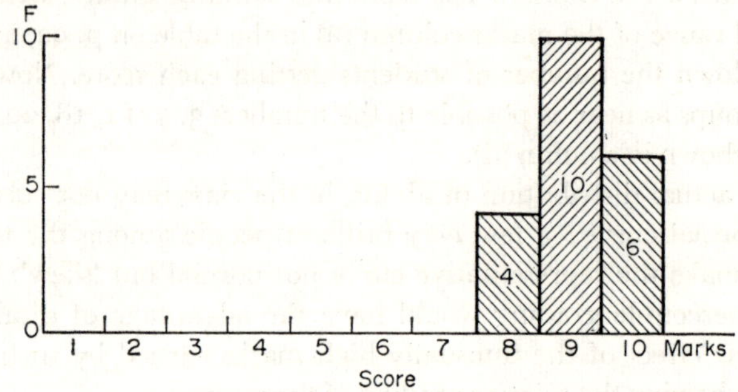

FIGURE 13 Result of too-easy test

This test was too easy in that it did not test all the class to their full capacity. Six students obtained full marks, half the class obtained nine marks, and four eight. The scoring range is only three marks and any attempt to grade the class would be of very doubtful value.

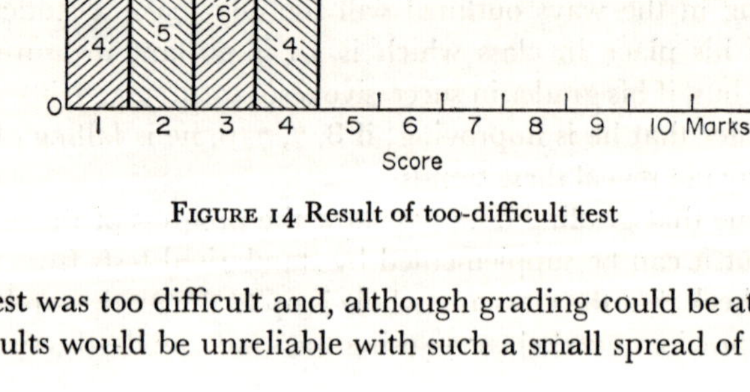

FIGURE 14 Result of too-difficult test

This test was too difficult and, although grading could be attempted, the results would be unreliable with such a small spread of marks.

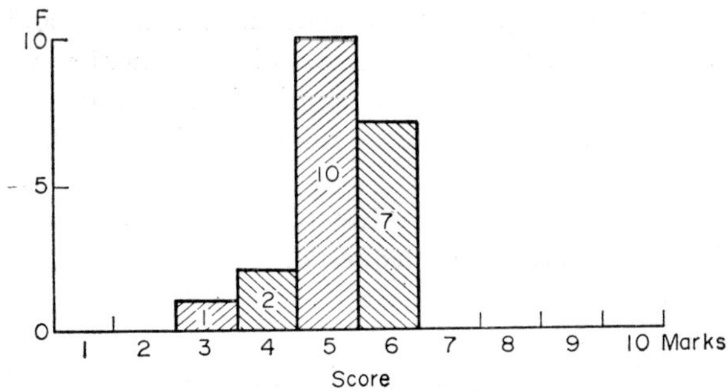

FIGURE 15 Result of badly-balanced test

This result might be obtained when the test contained a mixture of questions, some of which were too easy and some too hard and beyond the capacity of anyone in the class. All but two or three dullards get marks between 5 and 6. No reliable grading is therefore possible.

Here is an example of a well-devised test.

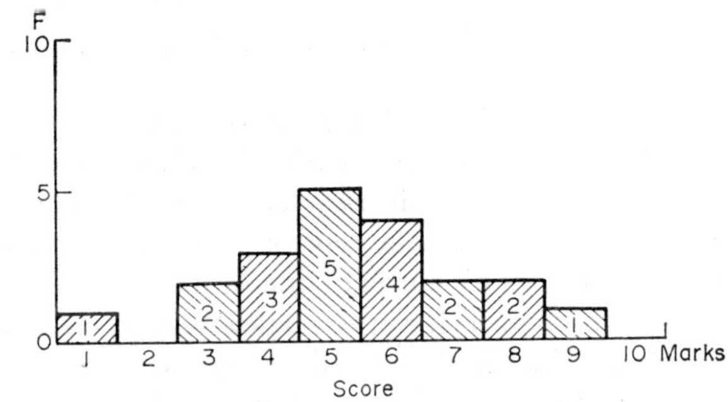

FIGURE 16 Result of well-devised test

The score extends over the whole range of the marks available and the class can be rated with a measure of confidence. The distribution is very roughly a normal one, but a better approximation to the distribution in figure 11 could be expected with a class of one hundred. With a good test and a class of twenty-five or so it is possible to give a stanine grading.

To be fair and effective, a test must have four main characteristics.

1. *A test must be valid.* This means that it must measure accurately the particular ability which it is intended to measure or test. It must have a clear purpose. For instance, in a test for technical knowledge,

a series of questions beginning 'give a description of . . .', and marked as an English essay, would not be a valid test; neither would one intended to test quality of workmanship, if measured mainly by a stopwatch. Also, a test of memory should not be regarded as a test of understanding principles.

Consider a simple example—two trainees in a test on technical calculations (two marks for each question):

Bill gets 10 right.

Fred gets 9 right.

Bill's work is very badly written down and he cannot spell, so the instructor takes off two marks. Fred's work is so well written out that the instructor gives him an extra mark for neatness.

The result therefore is Bill 18 marks, Fred 19 marks. Is this test valid? The instructor has, beside calculations, marked presentation. It is better to make this aspect a separate exercise. Mixing objectives often invalidates the test.

The instructor setting a test should keep the purpose clearly in mind, and avoid attempting to test two or three different kinds of ability at the same time. The marks obtained are valid for neither.

2. *A test must be reliable.* It is well known that examiners differ widely in their assessments, particularly if marking procedures are not clearly defined, as in certain essay-type tests. Further the same examiner marking the same essay type of paper after an interval of time may, depending on his current outlook, mark quite differently. An objective test is one which can be marked only in one way.

A test is reliable if it can be set to the same group and marked by different people at different times, and the candidates get roughly the same scores.

The test is thus consistent.

The basic condition for reliability is that it must be objective and therefore independent of the personal feelings, attitudes, likes or dislikes of the examiner.

It is fairly easy to devise a test of high reliability—so that there is consistency over the scores on repeating the test later, but a reliable test is of no value if it is not valid.

We may measure a quantity several times consistently, but it does not help if we do not know what we are measuring!

If, however, the reliability of a test is low, we cannot have much faith in its validity.

Some tests have both doubtful validity and reliability. Tests of mental aptitude and personality characteristic sometimes rapidly conducted at interviews may fall into this category.

3. *A test should be comprehensive in scope.* This means that it should cover the subject area fairly thoroughly. We all know the feeling of being unfairly treated when confronted by an examination paper that tests only a small fraction of what we have been working hard at, and encouraged to prepare! An examination consisting of one or two searching questions on a part of the syllabus may give an unfair advantage to a lucky student who happened to know that particular part, but it would spell disaster to the student who was unlucky enough to be examined on the one gap in his knowledge. Comprehensive tests put an end to the risky game of pre-examination question spotting.

4. *A test should be discriminating.* That is it should make it possible to differentiate between the best and the worst performers, and to grade the class by a good spread of marks from top to bottom. An inter-quartile range of 15 or more marks out of 100 in a reasonably 'normal' distribution will secure this condition. In simple terms the questions should not be too hard to cause a heavy concentration of marks at the bottom, or too easy to enable everyone to score high marks, as already discussed, p. 94.

In analysing the results of tests it is useful to check the way specific questions were answered—for feedback to the next test. To make the tests really discriminating, questions should be set which about 50 per cent of the class can get right.

A question which 5 per cent of students answered correctly is too hard to be useful and a question which 95 per cent answered correctly is too easy.

A way of measuring how discriminating a question has been is to take the top overall $\frac{1}{3}$ of trainees, group A, and the bottom overall $\frac{1}{3}$, group B.

Let percentage of A who got the question right be P_A
Let percentage of B who got the question right be P_B
Then $P_A - P_B =$ discriminant.
If $P_A = 90$ per cent
$P_B = 5$ per cent
discriminant $= 85$ which represents a good question.

If, however, $P_A = 50$ per cent
$P_B = 50$ per cent

Then the discriminant=0 and the questions would not be considered a good one, as it would seem to bring down good students or to upset both good and bad students. It might have been badly worded or it might have been a trick question.

Tests which are valid, reliable, comprehensive, discriminating and objective can be devised with care and practice by trained and experienced teachers. Books for further study are given in the bibliography.

In large training establishments such as those in the Services where often over 1 000 men are under instruction and have to be tested and assessed regularly, perhaps once a week, for progress records, upgrading, promotion and the validation of courses, the systematic setting of tests so as to be fair, reliable and valid is a matter of great importance and considerable complexity. There may be 100 classes under instruction, each has to be tested every week and each test requires 25 questions that cannot be anticipated even by shrewd guesswork.

Here the computer can come to the rescue, and operate with a 'bank' of say 25 000 objective and carefully validated questions. The computer is programmed to select questions for specific classes and purposes, and then deal with the results and the mathematics of grading and statistics. It can keep useful data and 'print out' on cards what questions have been set previously, for whom, when, and how often. It can make a full analysis of each question in the bank under the heads of facility factor or the

$$\frac{\%\text{ of trainees who answer the question correctly}}{\%\text{ of the trainees who attempt the question}}$$

and the discrimination factor which has been mentioned above.

Questions with a facility factor above 85 per cent or below 35 per cent are looked into and suitably modified, and questions with an excessively low discriminating factor are similarly subject to scrutiny and modification. Computerised systems promise to be great value to instructors, by producing a stable, reliable and valid system of testing—but the content of the tests and the devising of the objective questions still have to be done by expert training officers or teachers.

The best tests require a lot of careful preparation, for example a large number of short questions covering the whole subject area to be tested, the 'multiple choice' type of test already discussed, or a 'complete the statement' type of test. They are exacting for the instructor at the setting stage, but are easy to mark, valid, reliable and objective. Tests that are easy to set, for example 'briefly describe, etc.' are very difficult to mark.

Finally, instructions should be few, clear, and limited to essentials. It should be a reasonably economical exercise in time and energy both for the student who had to go through the ordeal, and for the teacher who has to mark it.

There are many different kinds of short objective test questions; some are mentioned below.

1. Multiple choice questions. Already discussed.

2. Complete the sentence. Normally a few key words are missing.

3. Single-answer question. Can be answered by a word, a phrase or a symbol.

4. Observation question. For example a diagram with parts A, B, C etc. shown which have to be named or their function explained.

5. Statement of steps in procedure. Answer required is a list not an essay.

6. Complete the diagram question.

Limitations of the objective tests

Having presented the case for objective test, it must be recognised that their validity is limited mainly to testing knowledge, understanding and, to some extent, attitudes; but they do not provide a means for measuring a trainee's powers of self-expression, his initiative, or ability to organise material in writing. These are often important abilities to be developed in some courses, and for testing these written examination papers of the essay type are likely to continue.

Technical training final tests generally are of two kinds according to their purpose.

1. The grading of trainees for progress reporting and the selection of men for promotion or transfer.

2. The 'pass or fail' objective for certification, e.g. a Trade Test.

1. This requires the kind of comprehensive test that will grade trainees in mark order. The actual marks used are only of importance as a means of effecting the grading. As no definite standard has been prescribed, the test need not be standardised, for example by Trade Union regulations, but it must be valid and reliable.

2. The 'pass or fail' test must determine whether or not the trainee has reached a defined standard of achievement in knowledge or skill. It is important that the standard should be professionally well defined for certification purposes.

Here objective standardised marking is important if the certificate is to be fair and widely accepted as a qualification.

Considerable unfairness can arise when these two purposes are mixed up as sometimes happens in selecting trainees for a higher course.

Continuous assessment

This is now becoming a useful alternative to the old anxiety-producing once-for-all examination or two or three crucial examinations but it has a variety of forms.

It can be a series of weekly tests, the results of which contribute to a trainee's final qualification. This kind of continuous assessment produces its own brand of continuous anxiety for trainees.

A more humane kind of continuous assessment has been tried, that is the assessment of practically everything that a trainee does including essays, practical work, notebooks, drawings, tests, reports and projects. Such an assessment calls for close co-operation with instructors and agreement on procedures.

This has produced perhaps more strain and restraints than it was designed to relieve, because of the constant demands for marking, and designing and producing markable work to a regular schedule. Some students may put themselves under a strain trying to win marks or approbation at all times.

In continuous assessment it is possible and desirable to give more weight to assessments carried out towards the end of a coure to overcome the generally poor performance of trainees in the early stages due to newness of the subject and getting settled.

The most satisfactory form of continuous assessment is that which conforms closely to a good learning situation, and which consists in frequently letting students know how they are getting on. Sound learning is quick feedback learning and this aspect of continuous assessment is the basis of individual methods of learning—by programmes.

It is now appreciated that there are many facets to the planning–teaching–learning–assessing feedback process as the lists on p. 213 attempt to show.

Each skill has a teaching method, a learning method and a means of assessment for optimum effect on behaviour and fairness in the final assessment.

Summary of practical points in devising, conducting and marking tests

1. It is generally more satisfactory and easier for marking to set a large number of short-answer questions, rather than a few questions that require long answers.

2. Questions should be clearly worded. Have a clear idea of the answer you expect for full marks when you set the question.

3. In the alternative-answer type of test, the answer sheet can be gridded and a stencil cut with the correct answers written on it, so that marking is simplified and is independent of the personal judgment of the marker.

4. A test of skill must be practical. Test accuracy first, speed later.

5. Test knowledge and understanding, not powers of expression. Set some task in which students may be permitted to use their notes or textbooks. The questions should be designed to test understanding of principles and ability to use them.

6. Good seating arrangements with candidates well spaced and no serious distractions.

7. Everything ready: paper, pencils, instruments, apparatus and instructors. To minimise any personal bias in marking, examination papers should not include trainee's names.

8. Be punctilious over the timing, starting and finishing, and indicate the passage of time at reasonable intervals. Allow no talking or communication between candidates under test.

9. Use a red pen or pencil for marking, as this helps the eye to pick up the marks more easily for totalling.

10. Give the results to the class the next day if possible and use a post-mortem to teach or confirm.

11. Decisions regarding a man's future should not be based on the results of tests alone, but tests help to overcome the unfairness in promotion by mere seniority or by a personal opinion.

12. When marking, the examiner must constantly be on guard against changing his standard as he goes along, and should rest from time to time as fatigue can lead to injustice. Whenever two or more instructors mark the same examination, they should confer both be-

fore and after marking to discuss standards and to correlate results.

13. The usual method of awarding marks is to attempt to assess the degree of proficiency with 100 as the top mark. It is applicable when a defined standard of knowledge or an objective score is involved.

14. For marking based on personal judgment, the criterion is generally pass or fail, and, if pass, whether honours, distinction or credit can be awarded. For this purpose letters may be used with say A very good, B good, C fair, D fail, E bad failure.

A 'chalkboard' or 'feltboard' summary of points made about examination and tests is given on p. 110.

Progress reporting and selecting men

Simple satisfactory tests cannot be devised to test a man's will to work, or his integrity, or his ability to co-operate with others. These qualities have to be estimated by observation over a period of several weeks or months.

One of the ancillary functions of training is to provide a means of assessing a trainee's ability and aptitude for a particular job, so that a grading of men by ability becomes possible, and a selection of men for advancement can be made with some confidence.

A full account of the methods of selecting and placing men is beyond the scope of this book, but instructors and supervisors of training should be acquainted with the general principles of the subject and should appreciate that good training can be wasted by training the wrong man. The round peg in the square hole produces frustration, inefficiency and unhappiness. He may work for half a lifetime under physical or psychological strain leading to conflict with fellow-workers, and breaches of discipline. In another type of work he might have been efficient and well adjusted.

It is important to find out by means of observation or a preliminary test before training what type of work a man can best do. If this cannot be done systematically, then early in a training course, men who are clearly unsuited to a particular type of work should be given a chance to change to a more suitable type.

Selection at all levels and at all stages must be made by a planned and systematic method and not left to blind chance or personal prejudice.

When considering whether a trainee should be promoted or demoted, a training officer should not rely on the result of a single examination, but should take into account a well-documented personal file containing all relevant information and educational history in the form of a cumulative record. This should supplement the analysis of all his tests and examinations, gradings in class, and progress reports.

In many jobs, qualities of temperament and character, such as reliability, integrity, initiative and self-confidence, are more important than the technical skill involved. A job of skill and interest for one man may become for another, equally skilful, a matter of routine and repetition. Some skilled men can accept repetition cheerfully, but others of a different temperament become restless and bored.

The training officer may be confronted with three types of selection : 1. recruiting and selecting new men from outside the works or company; 2. selecting within the works for training or promotion; 3. selecting for advancement on the results of a course of training.

In all three cases the problem has two main aspects :

1. What the job requires of the man;

2. What the man can best do with satisfaction to himself and to the firm. These aspects are now considered more fully.

1. *An analysis of the human qualities required* for success or failure in a particular job or type of work, is called *job specification*. Briefly this involves listing the physical, mental and social demands it will make on the trainee, for example does the work require any special physical standards? good eyesight? manual skill? (are left-handed people at a disadvantage?) ability to perform rapid calculations? ability to co-operate with others? ability to supervise others? a sound judgment? What are the 'essential qualifications'? What are the 'disqualifications'? What are the unattractive limitations of the job?

An analysis on these lines should be made for each job by observation and discussion with experienced and successful workers and checked with others.

It should be remembered that a successful performance in a job does not necessarily mean full personal satisfaction for the worker; if the job demands a good deal less from a worker than he is capable of doing, a good man may easily become very dissatisfied and eventually unsuitable.

2. *Assessment of the man in relation to the job.* The job specification largely determines what tests are to be applied and what questions can usefully be asked at the interview. Generally information is required about a trainee under the following headings : *physical capacity; intelligence; aptitudes; attainments; personality, including temperament and character.*

The selection of men by means of properly devised tests or a battery of tests, instead of by blind chance or wasteful trial and error methods, has had excellent results in the Services and in industry whenever they have been properly used. Rarely have men so selected been discharged or transferred for unsuitability. For example, in some works misfits have been reduced from 20 per cent to 5 per cent by using scientifically planned selection methods. Engineering firms are able to measure confidently a youth's technical suitability for a specialised branch of engineering after one hour's testing. An assessment of his capabilities by haphazard methods would be possible only after several months' work in the shops, probably accompanied by a good deal of strained personal relations or resentment.

The selection of apprentices for more responsible duties in special departments should be based upon a well-tried system of reporting and grading that operates throughout training or apprenticeship, and should include all relevant information recorded after regular tests, personal observation and occasional interviews. This grading and selection can be carried out with confidence, provided such systematic progress records are kept.

So much for the general principles. Now to return to progress reporting and selection of trainees by the training supervisor. Each trainee should have a dossier containing such papers as : (1) his application form, *vide* p. 1, chapter 1; (2) a job analysis to serve as a guide; (3) results of initial tests of all kinds, psychological tests, and tests and examinations taken under training; (4) periodical progress reports; (5) results of interviews.

Due credit should be given to trainees for the actual work done in class as distinct from tests, for example notebooks of practical work. Drawing and calculations should be inspected and marked regularly (continuous assessment) and the marks taken into account for the final assessment.

(1), (2) and (3) have been dealt with, and progress reporting in regard to learning and skills is largely a matter of recording results of tests and noting a man's grade, but reporting on a trainee's personality, temperament and character is vitally important, and should be done as objectively as possible.

In recording the results of tests of knowledge and skills we have seen that it is generally more satisfactory to place men in their order of merit than to attempt to give a marking by an absolute standard. In dealing with qualities of character and personality this principle is again applicable. Assessment of a person's qualities can best be done by considering how each deviates from the average trainee, or what the supervisor from his wide experience has in mind as the average trainee on that particular course.

An instructor's judgment is bound to be coloured by personal likes and dislikes, but this can be overcome to a large extent by considering each quality separately with conscious effort and detached deliberation, in much the same way as a wine-taster considers wine for body, dryness, bouquet and colour, each quite separately at first, before giving a final opinion.

A simple and effective way of reporting on a trainee is to list those qualities that are found by a job analysis to be relevant and important to his work, and then to allot marks on a five-point scale. This can be done in several ways, depending on which symbols are used to fix the average, and the deviations from it.

The five gradings are : + 2 outstanding (in the quality)
+ 1 above average
0 average
− 1 below average
− 2 seriously below average

The symbols A, B, C, D, E, can equally well be used, or the numbers 5, 4, 3, 2, 1, so long as C, and 3 are taken as the average and are not used to indicate a derogatory marking.

An example is given below showing some personal qualities that might be reported on. Each training officer should draw up his own special report form to suit the specific purpose of the training.

In conducting a course for instructors it is a good exercise in class participation to invite the instructors under training to suggest qualities they would wish to see in men working for them, or a list of qualities they think they might help to develop.

Attribute	*Marking*	*Remarks*
Technical proficiency		
Adaptability		
Loyalty		
Zeal and energy		
Ability to stand physical strain.		
Cheerfulness		
Reliability		
Punctuality		
Manner towards others or conduct in the works		
Tact		
Ability to co-operate with others.		

Reports are frequently regarded as an unwelcome chore and remarks are often of the kind that embellish school reports: 'must do better', 'should apply himself' and so on. They should indicate what the student must do to improve, and how he can best co-operate. Objective statements are more valuable than 'in our opinion', but even 'our opinion' (taken from various sources) is better than 'in my opinion' which may be based on limited access to all the facts and records.

A proper assessment of a trainee's capabilities as revealed during his course is of the utmost importance as it decides whether he is graded 'not qualified', 'qualified' or 'highly qualified'. This is in effect deciding where to draw two vertical lines on the graph in figure 12, to divide the area under the curve into the grades considered to have failed, passed or qualified for some distinction.

The interview as part of selection procedure

Selection procedures for people applying for courses or jobs usually depend on an essay type of test, numerous selection tests and an interview. It has been found as a result of several well-conducted studies of university students * that by far the best forecast of a man's ability is given by well-devised objective tests, and that a formal interview is of little value, unless conducted under suitable conditions, with all relevant information about each candidate available. In many selection tests and interviews associated with training the purpose is to select for trainability rather than for past achievement,

* See H. J. Eysenck, *Uses and Abuses of Psychology.*

but experience shows that, if a man has done well in the past, he is likely to do well after training in the future.

Most interviewers consider themselves to be endowed with a special gift for recognising within a few minutes the candidate likely to be a success. This results in subjective decisions being made or gives rise to disagreements within the selection committee.

To minimise the errors and unfairness attending interviews, those responsible should be aware of the difficulties and accept guidance from specialists in the subject. Interviewers tend to select the duller candidates and several people interviewing the same candidate frequently make quite different assessments. As a means of judging a person's character and ability, a short interview is often quite unreliable. An interview is important, however, in selections for jobs which require social qualities, such as the ability to get on well with other people.

Judge the man—not his ability to be interviewed.

One piece of advice the specialist gives—useful for instructors under any circumstances is contained in a BACIE Booklet *Interviewing in Twenty-Six steps* by J. S. Gough.

'Examine your prejudices. There is no scientific evidence to support the following: red hair goes with a quick temper. A high forehead signifies high intelligence. A firm handshake and square jaw denote a firm character. To look you straight in the face is a sign of honesty. (The 'confidence man' does this very well.) Agreement with everything you say means that he is a remarkably able person.'

Interviewing students about their work and progress

A valuable part of the training course should be a periodic interview with the training supervisor, or education officer.

An interview should be well conducted and impressive, particullarly if it is used to help decide which men are to be selected for more responsible posts.

The general pattern for an interview might be as follows.

The interviewer should study all the available reports and papers connected with the person to be interviewed.

Make sure that each student or applicant has proper notice of the interview, with adequate time allotted for each. Ensure that telephone calls and other disturbances do not interrupt interviews.

Trainees should not be kept waiting as this (apart from being discourteous) makes for nervousness. Put the man at his ease right from the start by means of a brief friendly conversation; this establishes contact.

Time must be found for proper interviewing. An appearance of haste or irritation at having to spend the time will defeat the purpose of the interview. Fifteen to twenty minutes is an average time. Encourage him to talk. Give him an opportunity to tell you about his personal worries if he has any. These often have a direct bearing on any failure at work.

It is important to establish contact with the trainee or applicant right at the start.

This is intended to gain the candidate's confidence, as the first few minutes of the interview is a critical period in his ordeal.

The interview should help the interviewer learn as much as possible about the candidate; outlook, attitudes, and fitness for a job or promotion. Candidates should be given a fair hearing. Interviewers should not monopolise the conversation, but rather guide it and encourage the candidate to talk and to ask questions.

In routine interviews by the training supervisor it is sound to tell the candidate how he is getting on early in the interview, as that is the question uppermost in his mind. Tell him also precisely what he must do to improve.

Give him some final and definite advice. It is useless to say to a man: 'You will make the grade if you put your back into it a bit more, or pull up your socks'; it is better to tell him that he should qualify if he spends three or four hours extra per week on his mathematics and pays more attention to his practical notebook.

Discuss his future plans, and leisure activities. Talk frankly about personality weaknesses that can be remedied.

Interview in a quiet relaxed atmosphere. If necessary, stagger the times so there is no queue outside the room.

The interviewer should end with some words of encouragement, no matter how unpromising the man may be. This ensures that he will give his optimum response to the advice given during the interview. Individual praise or blame makes a deep impression on a trainee especially when given at an impressive unhurried friendly interview, in an informal atmosphere. Here is an example :

Student: John Smith

Preparation:	Study original application form with all his particulars. Job analysis. Test results. Practical work, notebooks, progess reports. (Make necessary notes.)
Contact:	Rock-climbing and youth hostel interests. Recent performance in firm's band.
How he stands:	Should get a pass, and might get a credit if he can put in more time on machine drawing.
How he can improve:	Although good with hands and on practical work, should make a bigger effort over the theory of machines. (Show test results to bear this out.) A little more tact needed especially with some of the older men.
Personal affairs:	Mother's illness. (This explains the low marking for punctuality.) Opportunity to talk. Ambitions for the future.
Final encouragement:	You should make a good foreman in a few years. Keep up your good work with the band and good luck with the rock-climbing.

Assessment of training effectiveness

Validation tests

All training programmes should be subject to a series of controlled tests to ascertain to what extent they have been successful in achieving the objectives of the training. Such validation tests have two aspects. One is concerned with the extent to which a particular programme has imparted the actual knowledge, skills and attitudes that it was designed to do (internal validation). The other is concerned with the extent to which the programme has been successful in achieving wider objectives such as improvements in quality and quantity of production, reduction in accidents, and better communication and personal relations (external validation). See figure 41.

Management has to go further and make an evaluation of training —which seeks to measure the cost benefit of the training in social as well as financial terms. Management must be able to answer such questions as: 'Have the general and long-term training objectives of the organisation as well as the specific objectives been achieved, or are they being achieved?' 'Are the trainees able to carry out the full responsibilities of his job with competance and initiative?' 'Have they acquired not only new knowledge and skills, but also a new enthusiasm for their work?' 'Was it all worth the cost and time?' If not, then the training cannot be deemed a success. This evaluation is not easy to make but follow-up work on trainees and discussions with employers, foremen, supervisors, engineers, accountants and training officers will help. Instructors will find it helpful to keep notes of points arising in their work that are particularly useful or that could be improved upon. These notes can be conveniently kept in a loose-leaf book and might include material for lessons, interesting ideas or problems, and innovations appearing in technical journals, particularly those devoted to training.

Here is a summary of points made about assessment in a form that could be used as a transparency for overhead projection—revealing a section at a time. It could also be built up on a chalkboard, felt-board or magnetic board.

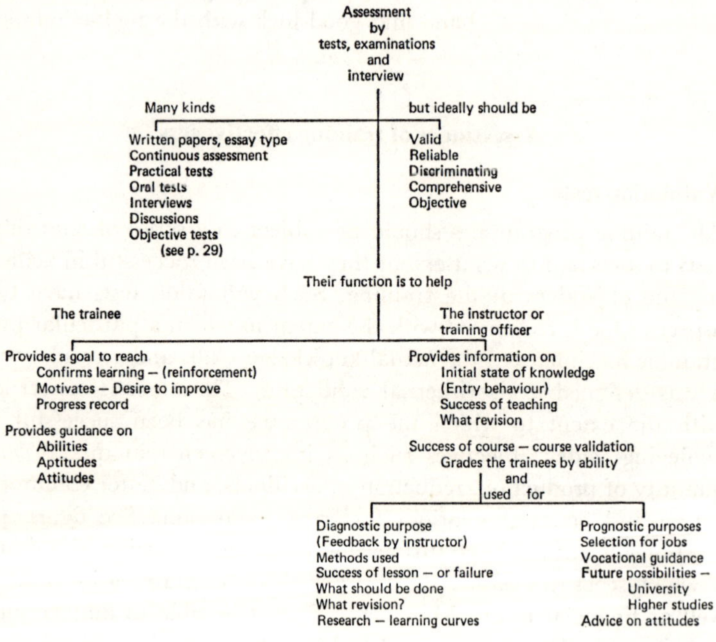

FIGURE 17

Feedback

Throughout this section the importance of *feedback* has been stressed because, no matter how well training is carried out and assessed, it cannot be fully effective unless the results of the assessment are fed back into the system when objectives are re-considered and new preparations made.

It is essential in all viable systems to have a sensitive feedback device. Without this, any system is liable to get out of control or stop.

The feedback may be positive and encourage the methods and ideas used, or it may help to suppress methods that have proved to be ineffective or inefficient.

Review of the Training Process

We have so far been considering the training process and various aspects of training under the headings of preparation, the instructor and teaching, the class and learning, assimilation of learning and assessment of learning.

We can now review the training function by means of a simple summary—see chalkboard summary (figure 18).

During a course or a talk this summary can be built up by means of feltboard or magnetic board.

A chalkboard summary of the training function

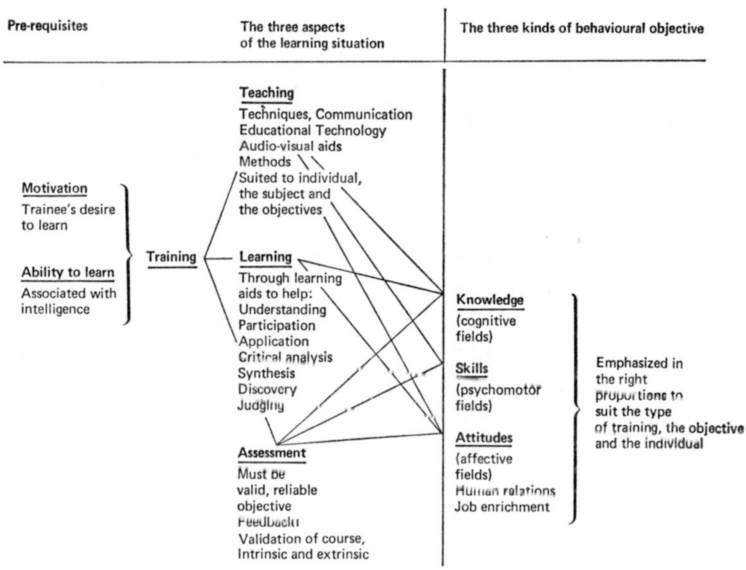

FIGURE 18

7

More about the Learning Process

THE principles of good training are based on the accumulated experience of successful teachers and on results that have been obtained from experimental work conducted on classes or individuals. Some people have a prejudice against anything which seems to be theoretical and will be heard to remark : 'It is all right in theory but how does in work out in practice.' It is as well, from time to time, to reverse this attitude and enquire : 'It's all right in practice, but how does it work out in theory?' Both attitudes have a proper function in the scientific method. We have seen (p. 60) that long practice by itself may not 'make perfect'. It is practice assessed, and with results fed back, that makes perfect. Theory and practice should not be regarded as being by their nature in conflict. Sound theory must be looked upon as that which will explain, guide and promote sound practice. If theory *does* seem to conflict with practice, then one or the other, or both, must be modified.

A knowledge of some of the experimentally established facts concerning instruction will often suggest to the inexperienced instructor ways of improving his performance.

Learning curves

One useful investigation is to find out how a class assimilates a subject as the course progresses. There are many ways of doing this. For example, the degree of proficiency attained at each stage, and hence progress, can be ascertained by several methods such as the following : 1. measuring the time taken to do a fixed task; 2. measuring the amount of work done in a fixed time or during a fixed task. A convenient way of recording the results is to plot a learning curve. The amount of assimilation, or the degree of proficiency in a skill is plotted vertically, and the passage of time, or the number of trials or the number of practices as the case may be, horizontally.

If a man can learn, say, six verses of poetry, or six pages of an engineering manual, in one day, it does not follow that in fifty days he will be able to learn three hundred verses or pages of the manual. Learning and time are not connected by a simple 'straight-line' relationship.

In an experiment by J. H. Bair on people learning to type without previous experience, each was asked to type a series of fifty-five letters three times daily for nine successive days. The results were as follows:

Day	1	2	3	4	5	6	7	8	9
Average time taken in seconds	79	72	63	60	56	54	53	49	47
Average number of errors	29	27	14	10	7	4	2·5	2	0·025

These results plotted on a graph are shown in figure 19.

The progress made is shown in two ways. The time taken and the number of errors diminish with each trial. (Note that the particular units used for plotting the curve determine whether it rises or falls as learning progresses. In this case, where time per trial and errors per trial are plotted vertically, improved learning or efficiency is shown

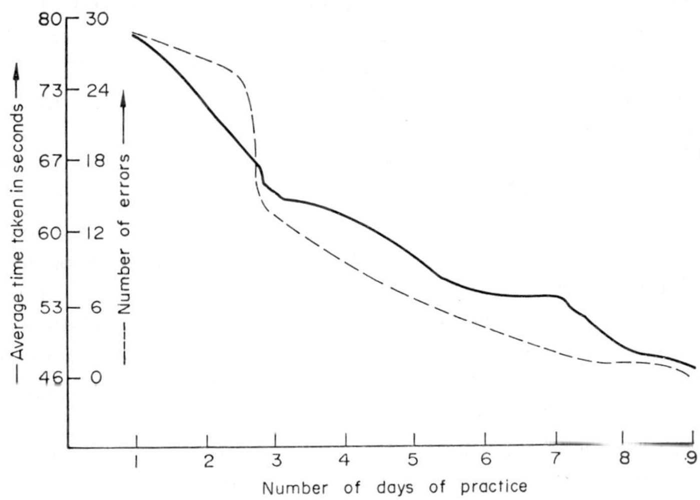

FIGURE 19 Learning to type: time and errors gradually diminishing (J. H. Bair)

by a downward trend.) It will be noticed that progress is much more rapid in the first three days than in the last three; this is a common experience.

A typical learning curve is of the form shown in figure 20. In this type, the proficiency achieved is shown along the vertical axis of the graph, and increasing efficiency is shown by a rising trend in the curve.

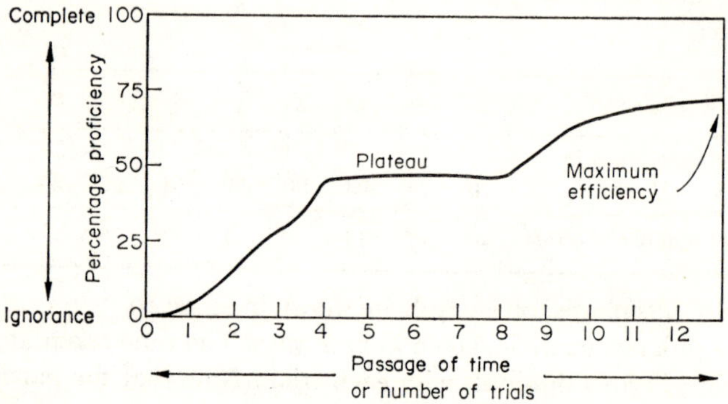

FIGURE 20 A typical learning curve

Plateaus

A steep rise in the curve means that the learning is rapid and when it flattens out the learning is slow. A flattening of the curve is called a plateau and in many subjects several plateaus can arise. When the curve flattens out finally and does not show any further upward trends, then maximum proficiency has been reached by the learner.

Plateaus may arise from several causes, among them being bad instruction, poor motivation, insufficient theory, insufficient practice, ill health, or even personal resentment against the instructor. Plateaus, however, can appear under normal circumstances or quite inevitably when the causes listed are absent. A plateau may mean that the learner has taken in sufficient for his mental assimilation (or muscular co-ordination which takes place in the brain and is similar to mental assimilation in the case of skill, and a period of time is necessary for further progress.

Plateaus mean that the student, whether we approve or not, is taking 'time off' to digest what he has taken in. They are a graphic warning against cramming as a substitute for understanding a subject. Cramming may succeed in giving a little slope to a plateau, but a longer plateau will probably set in later, with an even greater retarding effect on real progress and interest in the subject. Interest and enthusiasm rapidly die in a student if he feels he is not getting anywhere.

A typical example of a single individual learning to send and receive telegraphic messages in Morse code is shown in figure 22.

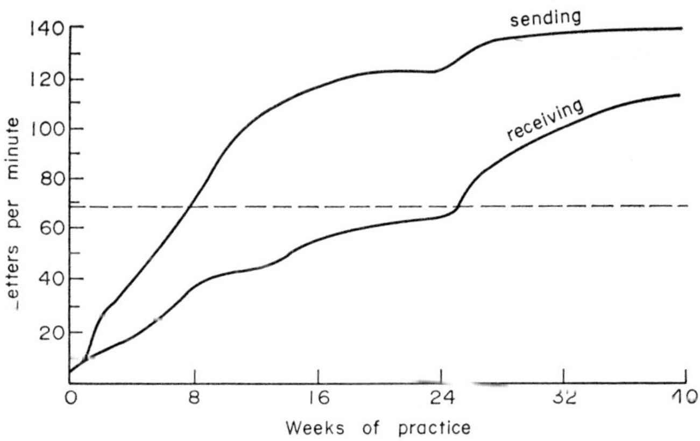

FIGURE 21 Learning curves: receiving and sending code during first forty weeks of practice

The curves show that proficiency in sending is easier to achieve than proficiency in receiving. This is mainly because the sender controls the speed, and his mental processes have advance warning of what is coming. He is able to give additional attention to difficult parts, and as he can work in complete words or even sentences, as opposed to single letters, his work has greater meaning. The receiver cannot do this and his interpretation and action lag as well behind the incoming signals. Note also that the receiving curve has a much more marked plateau commencing at about the tenth week than the sending curve. The wise instructor regards such periods of no progress as something to be normally encountered even under good training conditions, and he should explain to a class or a trainee in a state of stagnation what is happening and that a further burst of learning can be expected after a pause for assimilation and consolidation.

An interesting insight into another aspect of the learning process is provided by the learning curves (by Bryan and Harter in U.S.A.) shown in figure 21. These record the progress made by an individual learner who was tested at the end of each week following the sixth week of initial training in Morse code receiving. The curves show progress in receiving : 1. disconnected letters; 2. disconnected words; and 3. whole sentences.

From a very early stage, learning by sentences produced most letters per minute. After the twenty-fourth week the curve for receiving sentences rose sharply after a plateau commencing about the sixteenth week. This accelerated improvement seems to have arisen because the learner received whole phrases at a time and thus the phrase became his unit instead of the letter or word.

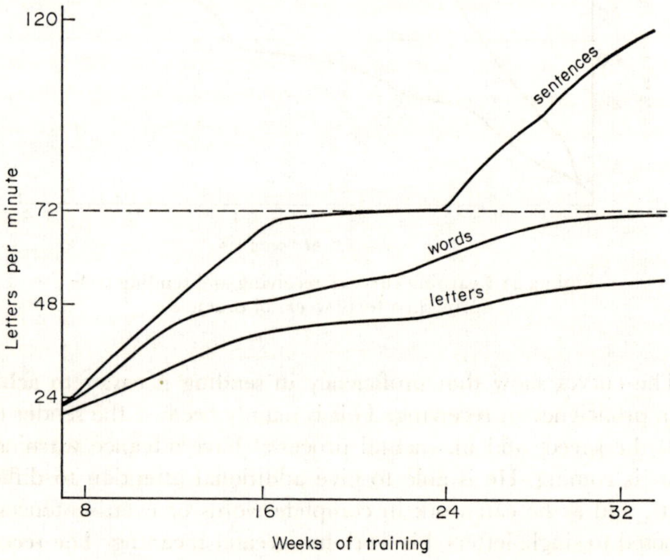

FIGURE 22 Learning curves : results of receiving Morse code in three different ways (Bryan and Harter)

Difficulty of the subject

Although individual differences make generalisations difficult, it is found that students learn a hard subject slowly at first, and make rapid progress when the principles have been understood and initial difficulties overcome. On the other hand, an easy subject is usually learned quickly at first but perfection achieved at a markedly diminished rate.

This principle can be applied in various ways to improve instruction. Easy skills such as filing can be taught with rapid progress at first, but perfection is achieved only after laborious practice. A difficult subject, say armature winding, needs much harder work on the part of the instructor at first, but, if the background skills are there, first-class work should come comparatively rapidly after the initial grasp of the work.

The instructor should explain to the class when approaching a difficult subject that progress may at first be slow, and that an apparently easy piece of work has a period of drudgery to be gone through before real efficiency can be reached. These advance warnings may save the class or individuals from a feeling of disappointment over their rates of progress.

We have seen that plateaus in learning curves can be caused by factors such as loss of interest or general decrease in motivation. Experiments have been conducted to test the effect on learning of various motivating factors.

1. *Letting a student know how he is getting on.* On p. 60 we saw that people work better if they know how they are progressing. Another experiment by W. F. Book and L. Norvell to illustrate this was conducted as follows.

Two groups of students, A and B, known to be of equivalent learning capacity were given a number of uninteresting tasks, for example crossing out given letters in lists of disconnected words in a foreign language, to be completed as quickly as possible. Each group was asked to do its best and worked for thirty periods, separated by suitable intervals. The members of Group A were allowed to count their scores, urged to compare them with previous ones, and encouraged to believe they could improve. Group B was given no such stimuli. Figure 23 gives the result in graphical form and shows that the motivating effect on Group A of knowing the results of its efforts approximately doubled the rate of learning in comparison with Group B.

2. *Explaining the importance of the work.* Two groups of students were each given ten problems of various kinds to be solved. Group A were told they were part of a grading examination following entrance to a college, and were important. Group B performed the tasks merely to fill in breaks in the course of other activities during their first day at college. Neither group were informed that they would be asked to recall their solutions later in the day or week.

Different samples sub-groups were tested for their ability to recall

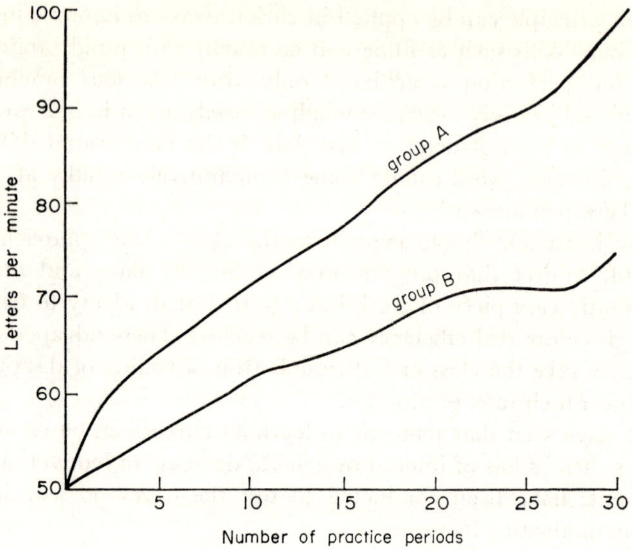

FIGURE 23 Learning curves: progress known and unknown (Book and Norvell) (see also p. 60)

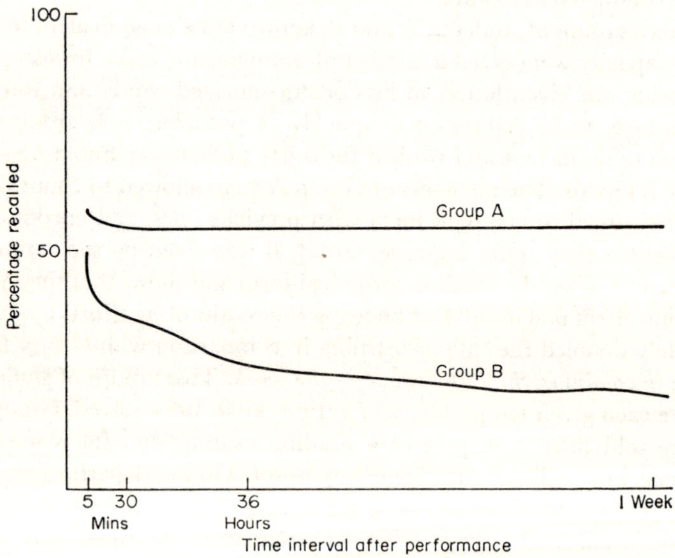

FIGURE 24 Curves of recall: improvement when task known to be important (W. L. Brown)

their solutions after the following intervals of time: five minutes, thirty minutes, thirty-six hours, and one week. The curves of recall in figure 24, by W. L. Brown, show how learning and recall are improved when the task is known to be real and important.

3. *The effect of praise and reproof.* (Experiment by E. B. Hurlock.) Four groups of senior schoolboys were given equal tasks: Group A was praised for all satisfactory work and given no reproof. Group B was reproved only. Group C was ignored and witnessed praise and reproof of others. Group D was ignored and isolated.

The results are shown in figure 25. Group A produced the best results, and the most sustained effort. Group B showed that reproof has some initial effect, but soon becomes ineffective. It had most effect on the best students.

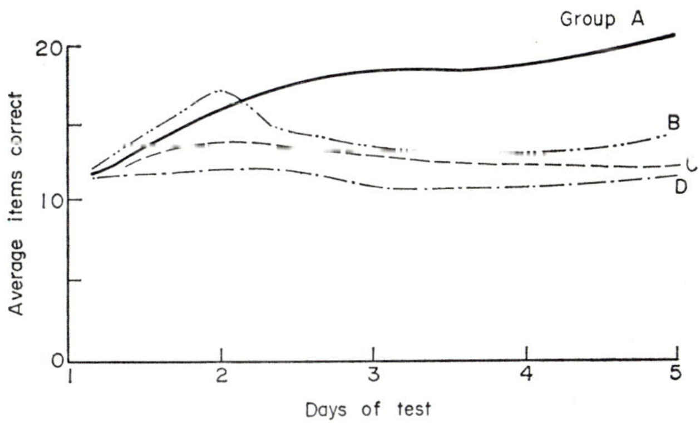

FIGURE 25 Effects of praise and reproof on groups of students (E. B. Hurlock)

4. *The effect of rivalry.* Three groups of average ability were set equal tasks. Group A was urged to do its best. Group B was split into two and one half competed against the other. Group C was divided into pairs and one member of a pair competed against the other.

The improvement achieved by each group after a suitable time was as follows:

A.	non-competitive	102 per cent improvement
B.	group rivalry	110 per cent "
C.	individual rivalry	157 per cent "

In this particular set of tasks individual rivalry proved to be a powerful motivation, but the dangers of overstressing this form of competition have been pointed out on p. 58.

From experience we all know that memory fades with the passage of time, that the fading is generally rapid at first, and then slows down. Recall prevents the fading. The memory is affected by numerous factors at the time of learning such as interest, emotion, associations pleasant or otherwise. Generalisation is impossible, but simple experiments in the learning of : 1. concepts and logical principles; 2. verse; 3. prose; and 4. nonsense syllables, give results that can be represented graphically (see figure 26). The learning in each case was only to the point where it could be reproduced correctly. It will be seen that after four days there is :

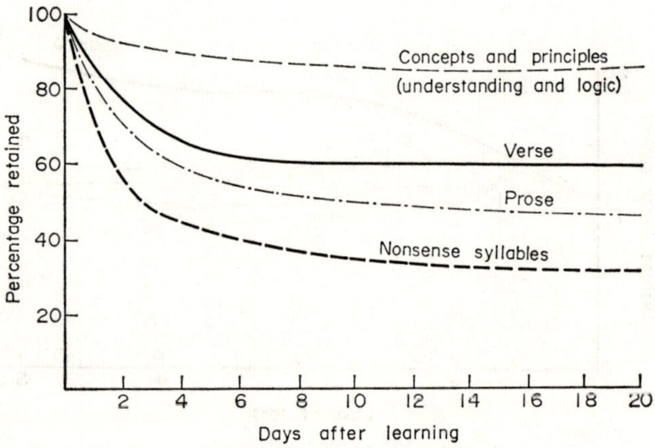

FIGURE 26 Curves of forgetting—for different types of material

(1) 10 per cent loss for the learners of concepts and principles;
(2) 25 per cent loss for the verse learners;
(3) 30 per cent loss for prose learners; and
(4) 40 per cent loss for those who learnt nonsense syllables.

These results emphasise the greater ease of recall of material that can be understood, particularly material that concerns logical principles; if you understand the basic principles of a subject, you remember it for a long time.

In each case the percentage retained drops rapidly in the first two or three days and then much less rapidly. A measure of the percentage retained can be arrived at by finding the number of repetitions required for full recall and subtracting this from the number of repetitions required for the initial learning.

Most of us have experienced the disastrous effect of *just* learning a piece of 'memory' work a few days before a test, and finding recall impossible at the crucial time. The graphs in figure 27 by W. C. F. Kreuger reveal the effect of: A. just learning to the point of recall say a dozen words in an hour; B. just learning and then overlearning by another thirty minutes; C. just learning and then spending an extra hour overlearning.

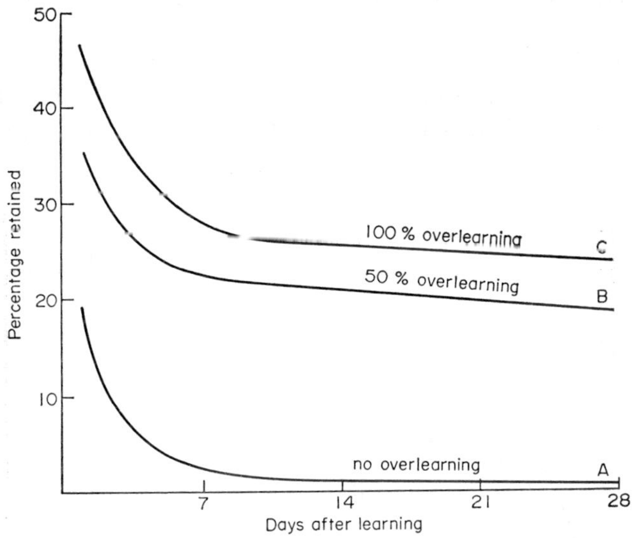

FIGURE 27 Learning curves: results of overlearning (W. C. F. Kreuger)

The conclusion is that, if the material is to be remembered permanently, the student must make an effort beyond that required for one recall immediately after learning. Instructors will find that devices for aiding recall are helpful, for example trainees should learn by participation, and instructors should use variety in method, emphasis and repetition of essentials, and timely revision.

Insight

In chapter 5 we saw that students of a difficult subject frequently made an advance by means of a flash of insight, which can be encouraged by the instructor in several ways. Experiments on animals learning by trial and error, and being rewarded with food, show how sudden insight gives rise to understanding of a situation, and no further mistakes are made.

The instructor should vary his approach to a difficult point and his method of presentation but should not show alarm or annoyance if the particular explanation which seems easy to him does not produce immediate enlightenment in all members of his class.

A great many technical subjects have points which are difficult to understand and much time can be saved if the instructor, after some experience, decides upon an effective method of teaching them.

When to test

The learning curve can sometimes indicate a suitable stage for a test or examination; consider figure 28. Position 1 is not a specially

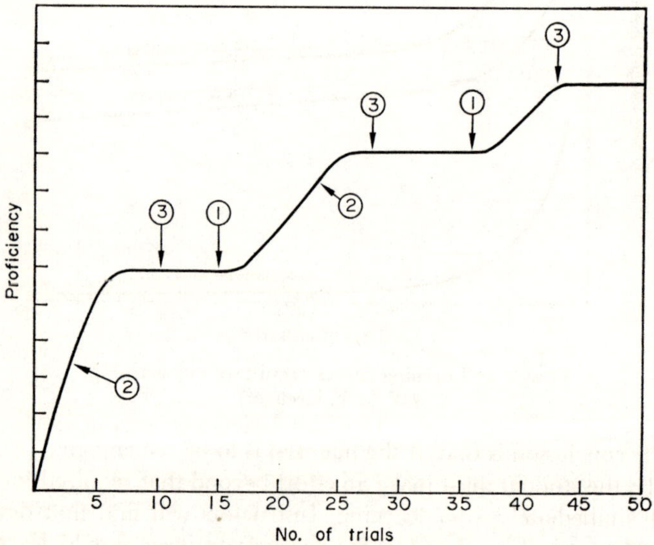

FIGURE 28 Tests and the learning curve

good time for a test because this point is at the end of a plateau and the class morale may be low, the students feeling that they are making no progress and that the subject is either difficult or boring. Position 2 is not ideal because here the learning is very rapid and by postponing the test a short time it would be possible to test much more. The best place is likely to be at 3 where a period of consolidation is about to set in.

Learning two or more subjects together

In all courses students have to learn more than one subject. It is important to realise that two subjects can help each other, or can interfere with each other and cause confusion in the mind of the learner. This confusion brings about inhibition. As as example a boy starting to learn Latin and Italian together may very easily become muddled in his vocabulary and grammer so that the two subjects inhibit good learning (see figure 29). On the other hand, English

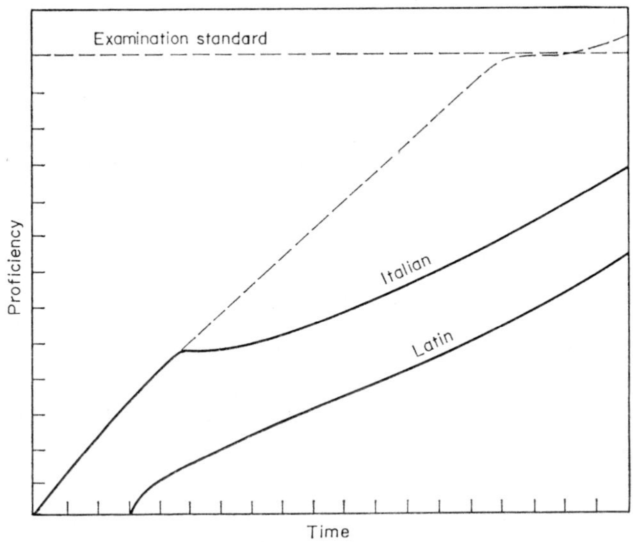

FIGURE 29 Learning curves: mutually inhibiting subjects

literature and English history could be taught so that they tend to reinforce each other. The boy might find English history rather dull and difficult to remember, and similarly English literature by itself might seem uninteresting, but, if taught side by side with references each to the other, the two subjects could be developed to build up a greater interest than would be found in either of them singly. This is shown in figure 30.

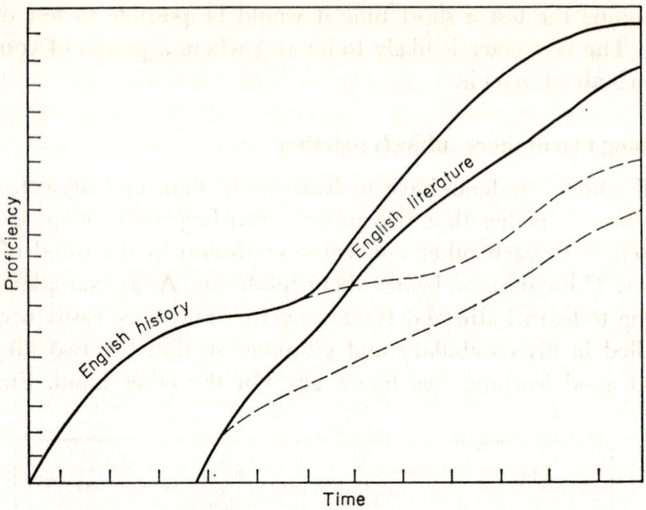

FIGURE 30 Learning curves: mutually reinforcing subjects

Applied mathematics and physics are subjects which, properly taught, can reinforce each other considerably. Figure 29 and figure 30 will illustrate these points. No set rule can be given for subjects that will inhibit or reinforce, but in general two or more similar subjects taught in the same course may cause inhibition whereas learning two or more aspects of the same general subject will produce a reinforcement. For example a.c. and d.c. machines taught in alternate lessons would almost certainly produce inhibition, but vector algebra and alternating currents could be studied side by side with advantage.

To diminish the effects of inhibition and to gain the advantage of two or more subjects reinforcing each other, it is necessary for training officers and instructors to arrange their programmes of work in full consultation and co-operation.

Saturation

A major hindrance to learning, that can be demonstrated by experiment, is that of 'saturation'. This is simply an overloading of the student with a mass of information on the same subject, particularly if the facts given have no logical connection. Saturation takes place when the material to be learnt is presented to students too fast for assimilation. It is no use presenting facts in a piece of instruction unless they are either useful in developing the subjects or likely to be memorised. In the humorous history book *1066 And All That*, only one date was acknowledged to be memorable and there is a certain amount of wisdom in this austerity over factual knowledge.

Saturation can be demonstrated as follows. Allow a class to listen to a factual talk on the radio for half an hour and three days later ask the class to write down all they can remember of the talk. A week or so later repeat the experiment with a comparable ten-minute talk on a similar subject and it will be found that slightly more of the ten-minute talk can be remembered than the half-an-hour talk. Saturation sets in when the learning is fast and continuous, and saturation point is reached sooner when material to be learnt consists of a series of similar facts. A fuller appreciation of this would improve many pieces of instruction; in fact disregard of the saturation point is perhaps the most common fault of all instruction (see also p. 21).

To show the serious effect of saturation, the following is an experiment (taken from the Army book *Good Instruction*) using a film heavily loaded with facts.

Two groups of men of the same average intelligence and age took part in this test. A thirty-minute film was selected which had as its aim to teach the organisation and duties of a certain type of work. The facts taught were of the sort that could easily be confused, for example the detailed organisation of similar departments.

1. Group A were shown the film right through. The film was then discussed for thirty minutes and summarised by the instructor. Next came a second showing with the instructor's final summary.

2. The same instructor showed the film to group B in three ten-minute sequences. After each sequence there was a discussion followed by a second showing of the same sequence followed in turn by a summary.

3. The same total time was thus expended on each method of presenting the film. Three days later a test on the subject-matter of the film was set for each class. The result was :

Group A : 28·8 per cent
Group B : 78·3 per cent

Investigations of the state of learning after each showing gave the following results. After the first showing a very large portion of the film appeared to make no impression on account of saturation. The beginning of the film was remembered apparently because the mind was fresh. The end of the film was remembered probably because the final impressions were reinforced by the relief afforded by the ending of the film. Even after a second showing of the film to group A the same saturation effect, for a large part of the middle of the film, was in evidence. This is shown diagramatically in figure 31 where the shaded portion corresponds to very poor reception, and attention caused by saturation.

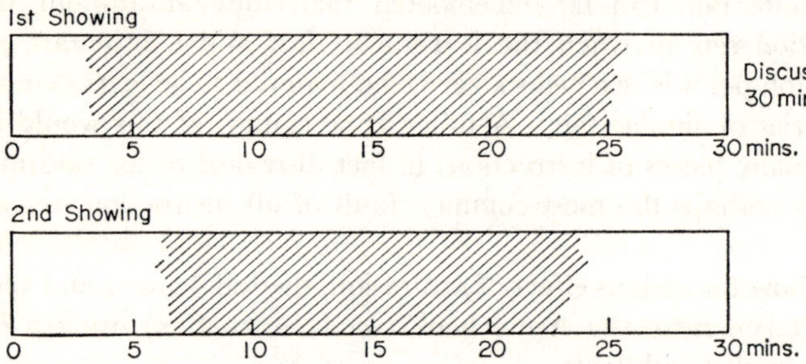

FIGURE 31 Class A: widespread saturation

In the case of Class B where the film was shown in ten-minute sequences, each sequence achieved approximately the same amount of useful learning as the first and last block of five minutes in the method used for Class A. A repetition of the method of showing practically eliminated the effect of saturation. Figure 32 illustrates this. The sum of the shaded portions from the three showings of ten-minutes, is considerably less than the shaded area from the second of the thirty-minute showings.

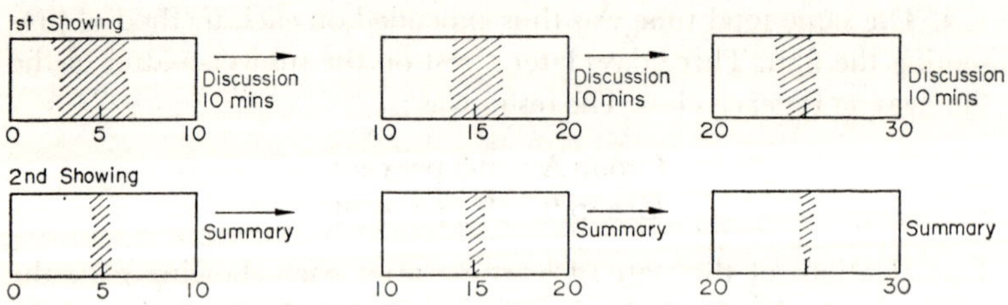

FIGURE 32 Class B: saturation practically eliminated

This experimental result has a wide application and, although it is unwise to generalise on this subject it is useful for the instructor to appreciate that most classes will reach a saturation point after about ten minutes of intensive passive instruction whether it be talking, demonstration or film-showing. After that period the learning must be confirmed either by means of a summary, by actual practice, by means of questioning or tests, or by 'giving back'.

A training officer who wishes to study in detail how learning is proceeding for a certain piece of instruction would be helped by plotting a suitable learning curve of the kind that have been discussed in this chapter, but to do this it is necessary to have a great many opportunities for collecting results, as conclusions drawn from a few samples will not be reliable. The test must be carried out with care on sufficiently large numbers and conditions must be unchanged for a series of tests. Full consideration must be given to any factor which might invalidate the results.

Care must be taken not to jump to general conclusions from learning curves for a certain skill by a particular group conducted by an individual instructor under a particular set of circumstances. Any attempt at a general conclusion can be made only after hundreds of investigations using 'control' groups, under the guidance of an expert statistician. This warning should not, however, prevent instructors from taking an interest in methods, rates of learning, measuring proficiency achieved and plotting curves.

Fatigue and efficiency

Technical training has to take into account the effects of sustained effort on a trainee, because exhaustion of physical or mental energy, known as fatigue, produces in general decreased efficiency and output. We have already seen a simple kind of fatigue curve, p. 19, which shows an initial rising output on the part of the class, followed first by a decline of interest and attention owing to a variety of factors including boredom, which is related to fatigue, and then by a brightening of the class at the end of the period. Fatigue is offset by the anticipation of the end of instruction. There is a common pattern for fatigue curves in workshops or factories for morning and afternoon sessions of work, and will be referred to later in the chapter.

The degree of fatigue experienced depends upon individual differences, and can vary considerably in the same person from day to day, and with different incentives or levels of motivation. Some people are prone to fatigue or its effects more readily than others, sometimes to a degree that can interfere seriously with the quality of their work, or output. Weary operatives may miss essential signals and tired drivers produce accidents. Fatigue leads to poor judgment, omission of detail, or indifference to essentials. Training carried out with the larger object in view can help to mitigate its effects.

Fatigue can be physical, mental, or a mixture of both. Prolonged mental exertion produces mental fatigue in a trainee; his eyes and face muscles feel tired; he will become irritable; and he will avoid brainwork and find it difficult to make decisions or to concentrate.

It is part of the function of training to help a trainee to build up stamina which will reduce the effects of fatigue.

Stamina building becomes a part of training in some kinds of work so that a trainee, who can already achieve an experienced worker's standard and speed and quality for a short run, can then carry on and maintain this output throughout the shift without experiencing any undue fatigue.

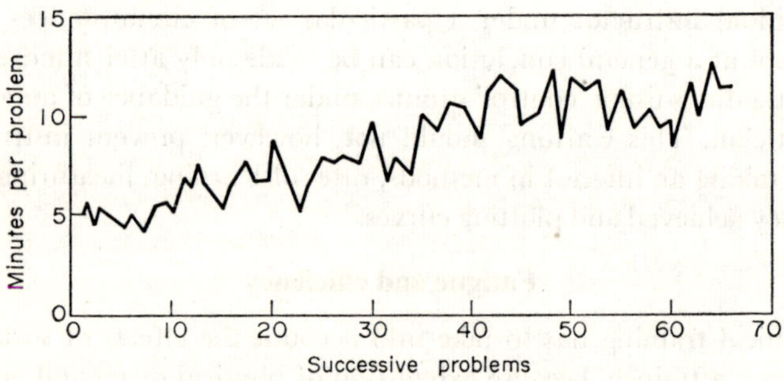

FIGURE 33 Mental fatigue: problems started early in day

Mental fatigue can be overcome by sleep or recreation including any activity which will hold his interest without making demands on his mental powers. Physical fatigue can be overcome by complete relaxation and sleep aided by suitable food and drink. The two performance graphs figures (33 and 34) are given as examples to illustrate how physical and mental fatigue affect a man's efficiency in performing mental tasks.

In figure 27 a man was given a number of arithmetical problems consisting of the multiplication of two sets of four-figure numbers. Starting at 11 a.m. he continued without a break until 11 p.m., completing sixty-seven problems. The time taken for each problem increased from five minutes for the first to about ten minutes for the last. This shows the steady effect of fatigue.

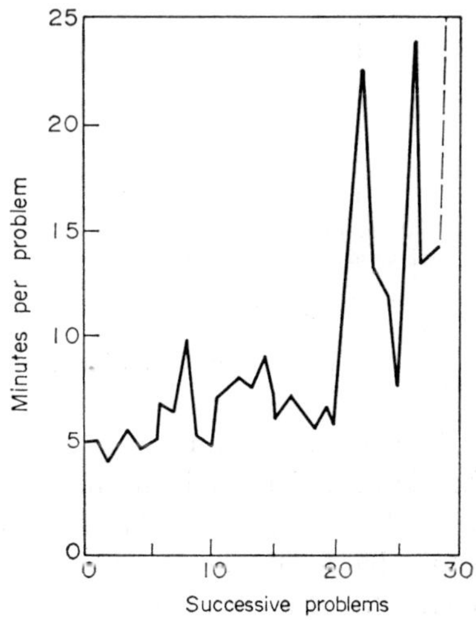

FIGURE 34 Mental fatigue: problems started at 11 p.m.

A different man was given the same test starting at 11 p.m. after a full day's work. For the first twenty problems his performance was much the same as the first man, as he took about one and a half hours. At this point his capacity for sustaining mental effort broke down completely as the next problem took up to twenty-five minutes and he was unable to continue (figure 34). Even making allowances for the different conditions and temperaments, these instances support the generally recognised fact that a man's vitality is at its highest during the middle of the day but is lowest at night. This should serve as a warning to students not to neglect sleep at night or not to work until the small hours of the morning. There is a natural rhythm in man's activity which it is hard to change.

Figure 35 illustrates how proneness to fatigue varies at different times of the day and night. Even when a man has been awake all night he will experience a renewal of energy round about dawn. The precise pattern of this rhythm varies with individual temperaments, with climatic conditions and with habits of life, but it nevertheless affects all men, so that, from the instructor's point of view, consideration should be given to this factor in planning a day's work.

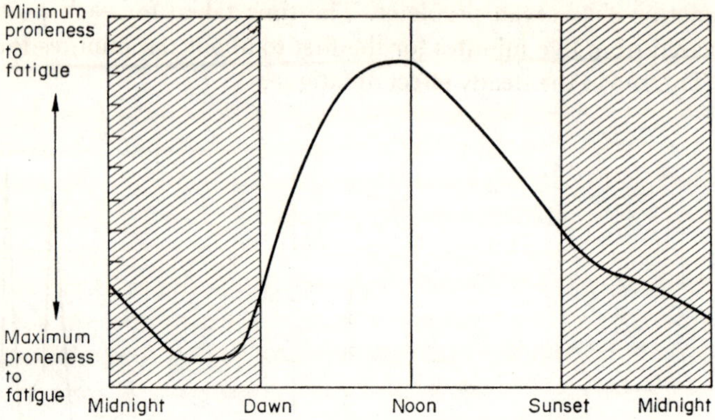

FIGURE 35 Variation of fatigue during 24 hours

The most exacting work of the day is best planned between breakfast and the midday meal. The more interesting work should come later in the day. Arduous mental work late at night is not as profitable for the learner as an equivalent time in the middle of the day. This rhythm in industry shows itself in the following typical work curve of a manual worker where fatigue can be regarded as steadily cumulative (figure 36).

Breaks offset the effects of fatigue

The efficiency lost by fatigue can, to a large extent, be counteracted by a wise use of breaks in the work. The initial rise in the graph shows that it takes a man some time to get settled in his work. It would be clear that too many breaks will produce a loss of time that

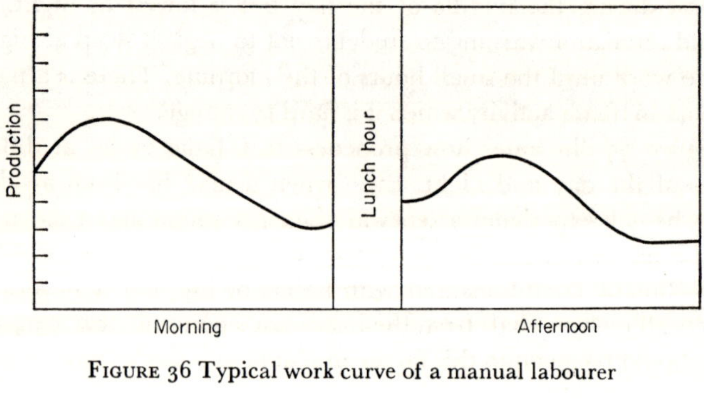

FIGURE 36 Typical work curve of a manual labourer

will counteract the gain from offsetting fatigue. It is more efficient to remedy fatigue before it becomes too acute; a short rest can restore full efficiency if the fatigue is not great.

In figure 36 the best position for a work-break would be just after the peak of morning and afternoon production. The results of introducing this work-break at these times are not quite what would be expected, as psychological factors play a part. For instance the anticipation of a work-break may cause a slow start but produces an increased out-put before the break arrives, as shown by the dotted line in figure 37. This tendency is again apparent in the second part of the morning, but the man on the whole worked more efficiently. In the afternoon the break seemed more distant and he worked well, falling off just before a break was due, possibly because he expected

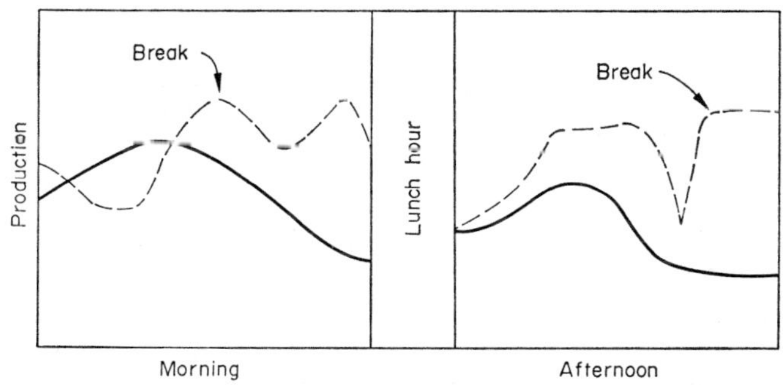

FIGURE 37 Work curve: result of introducing breaks

it sooner. On the whole he finished off the day's work more efficiently than at any other time. Studies such as these show the value of well-planned work-breaks.

The instructional period

For a class under intensive instruction there is no simple method of estimating the degree of fatigue of the class, but it is generally true that a period of forty-five minutes' instruction is the maximum that can be reasonably endured without a break or without serious loss of efficiency.

Experiments have shown that on the whole 'spaced time' is better than 'massed time' in the learning process. That is if you have two hours of instructional time to cover a difficult part of the training syllabus it is generally better to have four half-hour sessions than to have one two-hour session, even if the periods are separated by several days.

The breaks allow time for assimilation to take place. Consciously or subconsciously perhaps the learner has time to reflect on and consider what has been taught, and this is an important part of the learning process.

There are many parallels between the human brain and modern computers. Each has to have time for information to be sorted and stored ready for use, and it is possible to get confused outputs if overloading takes place.

Unfortunately trainees are often obliged by problems of staffing, accommodation or time-table considerations to attend continuous sessions on one subject for two or three hours at a stretch. This may happen on day release or evening classes, but it is not ideal for trainees or staff. Much can be done to mitigate the harmful effects by planning adequate breaks and changes in the subject matter interspersed with class and individual activity.

When the work is really strenuous, a break should be spent in proper relaxation, sitting down, and should not be too long or muscles will become stiff; and it should not be too short or the rhythm of the work will be interrupted—with only a small check in the onset of fatigue.

A production plant in the United States of America introduced a fifteen-minute work-break in the middle of a two-hour period of work. When the break was spent in complete relaxation, chairs being provided, production went up 9·3 per cent. When it was spent in walking about the increase observed was only 1·5 per cent.

Having seen the importance of work-breaks, it is essential in any training time-table for instructors or technical teachers to observe strictly the time-table and not encroach on these breaks as this is unfair for the next instructor or for the next part of the programme. In some programmes it is advisable to have short breaks every forty-five minutes with a half-hour break in the middle, but under such circumstances the half-hour break should not come in the middle of a piece of practical work, otherwise the trainees will find it difficult to start again at the point where they left off. This applies particularly to practice in skills and techniques.

Learning should be assimilated to a certain standard before it is safe to break off; then the break should be long enough to overcome

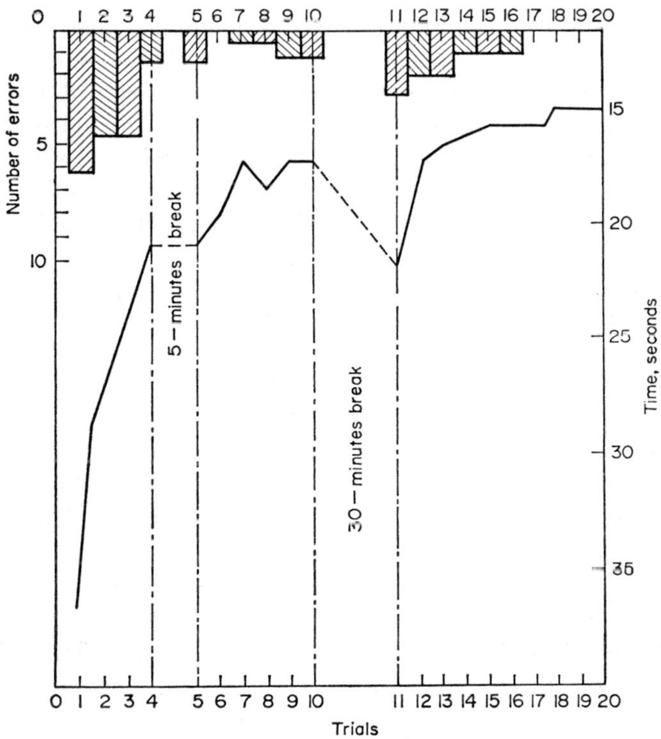

FIGURE 38 Breaks in continuity of learning

the onset of fatigue, but not long enough to allow learning to be lost. Figure 38 shows an example taken from a test on ten men learning a skill for the first time (the test was changing a part on an intricate piece of mechanism). During the first four trials learning progressed rapidly. After a break of five minutes no accuracy was lost and only a little speed was sacrificed. At the sixth trial maximum accuracy was reached, but then fatigue, and possibly boredom, set in. After the tenth trial a half-hour break was given. On recommencing the trials accuracy was found to have badly deteriorated; it took seven more trials before the old maximum was reached and another two trials before there was any improvement.

In chapter 5 we have seen the importance of variety in maintaining interest. Variety in occupation and in the training programme serves much the same purpose as rest; in fact, 'a change is as good as a rest'. This is generally taken into account as a matter of common sense in drawing up a training time-table. For instance, use might be made of a film, or by using discussions in place of lectures, visits outside the class-room, and practical activity interspersed with theoretical work. A final warning : no matter how well the time-table is arranged, no amount of artistry in arranging breaks and variety in programme can offset the results of accumulative fatigue if the hours of work are too long, and the intensity or tempo of the work is too high.

In a factory working at great pressure it was found that by reducing the working week from fifty-eight to fifty hours, the weekly output was actually increased by about 20 per cent. In a long course of training this principle, together with that of saturation, should warn the training supervisor to guard against fatigue and staleness, and to take measures to arrest it at an early stage. It is important to remember that fatigue can be overcome by an effort of will. Fatigue is not like a drug which cannot be struggled against, but is intimately associated with will-power, and motivation.

An explorer on the point of making an important discovery or a scientist about to unravel a new mystery may be motivated to work harder and faster in spite of considerable fatigue. In brief, motivation offsets fatigue. Figure 39 shows an attempt at measuring the effect of fatigue on men when motivated and when bored.

A class was set a series of well-devised intelligence tests and later a series of addition sums; as the work progressed their output was recorded, and they marked down the fatigue they felt on a ten-point scale from 1—'no fatigue at all' to 10—'too tired to work'.

The class found the intelligence tests interesting and their output consequently increased, in spite of a steadily increasing fatigue. The task of addition proved to be dull and this meant poor motivation and a steadily decreasing output as fatigue increased.

In arranging a programme, it is often preferable to ensure that men are well motivated rather than to ensure they are not fatigued. Motivation can offset fatigue and it is more profitable to instruct a man when he is tired and interested than when he is fresh and bored. Men should be trained to withstand normal fatigue, as a man who gives up at the onset of a feeling of fatigue is useless in any worthwhile job. Opportunities for strenuous work under difficulties which produce fatigue can be provided as a part of training,

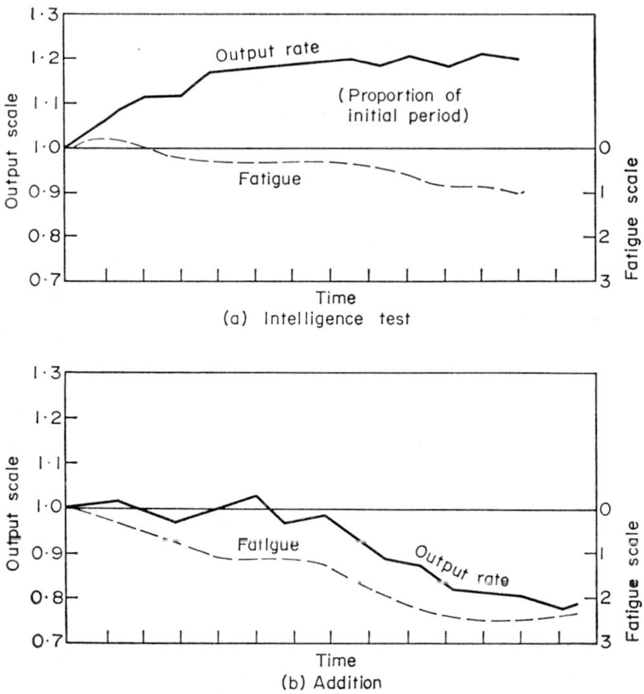

FIGURE 39 Fatigue offset by interest and motivation

but it is a piece of character training as distinct from technical teaching and calls for special qualities of tact and leadership on the part of the instructor.

The pace of instruction—tempo

An instructor will find that the pace and interest of his programme will vary from hour to hour, from day to day, and from week to week. Some parts of the work are full of interest, stir up enthusiasm and require a great output of energy; other parts are dull and seem to drag. Keenness and enthusiasm will naturally fluctuate because it is impossible for a man to be in a state of intense interest or enthusiasm for a prolonged period. The most enthusiastic musician could not listen to music demanding a high state of aesthetic interest for hours on end.

The word tempo is applied to the pace or degree of interest and energy demanded for a piece of instruction. It is impossible to make any rules governing the way in which tempo should be increased or decreased during a lesson or a course, but a study of this question has produced a few general principles.

A lesson which requires a short period of great activity should start at full speed and continue at high pressure to a central climax. Some loss of tempo in the second half is inevitable, but the lesson should culminate in a final burst at maximum pressure.

In teaching a new and exacting skill which necessitates revising several old ones, one programme or lesson plan could be as follows :

verbal summary of previous lessons	6	minutes
individual practice for revision	12	"
explanation of new skill	5	"
demonstration	5	"
individual practice	12	"
	40	"

The verbal summary calls for little class activity. Practice in the old skills increases the tempo a little, but it is likely to decrease by the end of ten minutes. The lesson ends fairly quietly with no marked enthusiasm or final burst at the end.

A second plan might be thus :

practical test as a revision exercise	6	minutes
explanation of new skill	4	"
demonstration of new skill	4	"
individual practice	7	"
competitive practice	4	"
practice revision preparatory to final competitive test	10	"
final competitive test	5	"
	40	"

The second lesson plan has a more sustained tempo particularly at the beginning and end, and calls for more preparation, and special enthusiasm on the part of the instructor. Note that after explanation and demonstration, the class proceeds to practice the new skill, motivated by the anticipation of a competitive test which follows the practice. The tempo drops during the revision and practice, but rises to a peak with a final set of competitive tests. It is important to note also that the tempo of the period depends not only on the planning and arrangement of the lesson, but also on the enthusiasm and manner of the instructor.

With regard to the day's programme we have seen that work which demands the greatest effort is best fitted in between breakfast and lunch. When students are tired or in an afternoon frame of mind, periods which demand little physical and mental energy are advisable, but for testing purposes a student should be called upon to produce high activity at any time. The tempo of each day must be given separate consideration, taking into account the subjects being taught and the larger cycle of training. In many cases the tempo of a day's work could be improved, as shown in these examples from army training where the second plan is clearly more effective than the first.

The instructor should strive to work the class at its maximum tempo for the type of work and time of day. Men on a technical course should be trained to work as quickly as possible and at maximum efficiency, so that the training is related to the needs of industry and leads to greater productivity and improved stamina.

Group structure

Our social system consists of an intermingling of numerous groups; the family, the club, the office, the firm. A man's character is not formed solely from within himself but is determined to some extent by the group of which he is a member; he will tend to identify himself with the ideals and thoughts of the group and he will be influenced by his relations with the other members of it. Personalities within any particular group become mutually adjusted each to the other, and this makes it possible to recognise certain particular characteristics of, for example, the only child, a successful business man, a schoolmaster or a naval officer. This network of personal relationships forms a 'group structure'. A class under training is a small community with its own group structure.

The farmyard gives a simple example of this structure in a very rudimentary form. If a number of hens are put in a run together for a month or two they rapidly become a social group with a simple structure of which a 'pecking order' is one of the characteristics. Hen 1 will peck all other hens but is rarely pecked back. Hen 2 will peck all but hen 1 and will enjoy immunity from all but her, and so on until the last hen is pecked by all without any apparent right to peck back. Individual characteristics become apparent. Hen 1 appears full of confidence and poise. The last hens in the 'pecking order' are maladjusted, vicious and a source of trouble. Keeping the hens in a group, however, is on the whole more satisfactory than keeping them alone. Egg production by the group is greater than that for the same number of hens kept in isolation. Men work much better in groups than as individuals. Although our present civilisation has been achieved by the constant striving of men to get the better of each other by socially accepted methods, yet the real secret of human progress is the fact that man can be unselfish, and work as a member of a family team or a community. As seen in chapter 4 both competition and co-operation play important parts in motivating people.

One of the characteristics of a flourishing community is that the members are good mixers. A few of them brought together with some common purpose will soon weld themselves into a group with a leader and some kind of organisation. With the increasing affluence of modern youth, giving rise to leisure, easy communication by motorcycle, car or telephone, groups or gangs of young people are easily formed and often seem to have no special aim except the seeking of recognition as a group, frequently by the challenging of authority in various ways. Training officers have to take into account modern social developments, particularly as they affect the outlook of youth. In the current idiom you must be 'with it', or at least have the group with you and understand students' attitudes and points of view. In class management problems be sure the group leaders are with you, and that there is mutual confidence between you and them. Treat all members of your class as important and responsible people. Group discussions can help here : let the members have their say in any problem; they will discuss it frankly outside anyway, without your being there to give guidance. A group rapidly develops a spirit which has been given such names as 'esprit de corps' or group morale. The Boy Scout movement makes excellent use of it in the patrol system which encourages the gang loyalty of boys which is a great aid to character-training. Individuals in a group behave differently from individuals on their own. A firm or a large organisation, if the group structure is sound, will be able to recognise the following gradation in the structure :

me;
Bill, Bert, and me;
our class, our shop;
our department;
our firm;
our industry.

Groups are generally from three up to about seven men and each group will usually be centred round one or two dominant personalities. These men largely determine the tone and standard of the group. Under the influence of good dominant personalities the group's performance will be higher than the sum of each individual performance; under bad influence—that is, one which is contrary to the best interests of the firm—the group will be a disruptive element or a subversive clique which can readily give a wrong lead to another large group. It is advisable under these circumstances to break up such a group. A sound group structure can encourage loyalty and good personal relations and should be encouraged to mature. There is a vast difference between an expert team of footballers when first brought together from different parts of the country, and the same team after it has played together for a season. The practical point of all this for the instructor is that a group of men should be recognised as a natural unit and that men should not be changed from class to class or group to group haphazardly but only for special reasons. During training and teaching, use small groups of friends as units in competitions as this develops a healthy group structure which ultimately promotes loyalty to the firm. When a number of men are presented for training and have to be divided into two classes, there is a temptation to select the two classes alphabetically or by some quite arbitrary method. The most effective way of arranging for the class is to grade the men according to their ability, that is, put the best men in one class, and the rest in the other class. Care should be taken that the classes are called by letters which will not obviously indicate which is the better class, otherwise group morale of the lower class may suffer. By grading classes in this way a more satisfactory group structure is secured and the learning is more efficient. Competition within each class is keener. Good instruction is easier and, as the speed of assimilation is more uniform in each class, time is saved. A large gap in ability in performance between the best and the worst men is a serious problem for the instructor. Experience shows that specially brilliant students tend to depress the worst students and not to inspire them.

There are, however, some disadvantages in dividing a class by ability. The different abilities may call for a modified syllabus and all members of the class will not reach the same standard together, but this disadvantage will be offset by each student being given the best conditions for assimilating fully what is taught.

Factory life in this half of the twentieth century offers some compensation for the drudgery of repetition work; it provides a social environment and a sense of belonging to a team, and a community. The old community life of the village and the parish, with the church at the centre of this life no longer exists in large industrial towns. Workers often live in dreary suburbs and look to the firm for their sport, recreation, club life and fellowship.

Social relations in a large firm are today complex and influential —for good or ill. So that in the words of Sir George Shuster, 'We stand at the beginning of a social revolution'. Training in industry must be sensitive to and responsive to these changes.

Sensitivity training

This is a comparatively new term which has come into use in recent years to describe an important aspect of the preparation of instructors, supervisors or managers for their jobs.

Like many new terms in the training vocabulary it concerns a process we are all aware of and which basically means a training in understanding and appreciating the needs of others, with a sensitive 'do-as-you-would-be-done-by' approach.

There is nothing new in this even as a part of an industrial training programme. The Training Within Industry programme 25 years ago appreciated the importance of human relations training for foremen, supervisors and instructors, as well as for managers.

We all receive some sort of sensitivity training throughout life, often not very effectively or pleasantly, from our day to day experiences. Training in human relations has always been an important part of the preparation of leaders, priests and social workers.

What is new is a fresh emphasis, and a fuller appreciation of its importance in achieving better job relations, job enrichment, job satisfaction, and the application of a systematic approach using modern analytical methods to achieve these objectives.

For instructors, what is required is not only systematic programmes or seminars with special talks, discussions or films dealing with sensitivity and ways to help people overcome the stress and strain of modern industrial life, but also for all training programmes to pay particular attention to the development of good relationships between individuals and among groups within organisations.

In such programmes, the emphasis would be on the development of a more responsive sensitivity to the needs of others and their problems, both personal and at work. This would entail better communication and co-operation between trainees and employees and employers at all levels, and would lead to improvements in job satisfaction and group performance. Sensitivity training for instructors will put proper emphasis on the learning situation, in which learning is recognised as the resposibility of the learner. This means that the learner's feelings, ambitions and frustrations are of great importance and should be understood and sympathetically taken into account by the instructor.

Planning a syllabus

In the Services, where training has always been taken seriously, it used to be the practice, now abandoned, to commence a training course by polishing up the trainees' arithmetical knowledge. Primary school curriculae used to consist largely of boring and exacting arithmetic and grammar exercises, because this had been the long-standing educational tradition, based on the mistaken belief that these exercises trained the mind and that this training would automatically be transferred to *any* subject subsequently studied. Psychologists have deflated this theory of 'carry over'. Training in subject *x* does not necessarily help subject *y* unless *x* and *y* have elements in common and the trainee has been taught and encouraged to recognise and use the connections.

A syllabus is the scaffolding around which training is built. The first step in planning a syllabus is to define clearly the *object* of the course. The object of the training may be laid down in a directive from management, or it may be largely dictated by the requirements of a recognised examination. See p. 152.

The senior instructor planning the syllabus must keep this object constantly before him and he should make a detailed study of all the requirements. The same is true when a syllabus needs revision, but here the adequacy or otherwise of the existing syllabus can be assessed by inquiring how far it has succeeded in achieving its object. This inquiry should be directed both to the departments that receive the men trained and also to the trainees themselves.

Suppose we have trained twelve men for ten weeks to be wiremen, we should like to know what weaknesses the receiving department found in the training. This may entail a systematic inquiry guided by a job analysis. This may reveal, for example :

principles of testing very weak in all trainees;
installing metal conduits good in most;
wiring regulations not well grasped.

Then we could ask individually a selection of the men trained: 'You have been on this wiring job now for two months. Did you find the course gave you the right start? How could we make the course more useful?' See p. 109.

A good training department will maintain close contact both with those who decide what the object of the training shall be—what is required—and with the finished product of the training course, i.e. the trained men at work. Without this close contact and co-operation, instruction will be unrealistic and ineffective. Instructors exist to supply the needs of industry, and they must be fully aware of those needs.

The results of training, assessed fully on the job can thus be fed back to benefit future instructors. The results of this follow-up work should be used to supplement the feed-back discussed at the end of chapter 6.

Having clarified the object of a course, the next step is to decide on the main subjects to be taught, and these have to be broken down into sections. The compilation of a detailed description of the requirements for the job is valuable for the instructor and often revealing to management, as the exact requirements for even a low-graded job may prove to be quite impressive to officials who have not given such matters much thought.

A detailed job analysis for all jobs will help to check any overlapping and will suggest common background subjects in training, for example, the history of the firm and its products. The training supervisor should examine these ingredients of the syllabus and decide how each part contributes to the main object and how important that contribution is.

In a technical training establishment the content of the syllabus and much of its planning is determined by the requirements of external examinations. The larger problem is the arranging of the whole time-table to ensure the maximum and effective use of staff, classrooms, workshops, at the best times of day for classes of the right size.

This is often done as a team effort on large time-table boards which can be up to 1·80 m. × 1·50 m. They can be divided into the days of the week, and various arrangements can readily be manipulated by using sets of cards with staff names, classes and rooms. Eventually, after agreement, the various convenient forms of the time-table can be copied, for example staff time-table, room allocation and subject time-table.

In preparing a lesson, it was suggested that material for presentation should be sorted into *must* know, *should* know and *could* know. The same procedure applies to the planning of a syllabus.

The next step is to consider the time available and the receptive ability of the class, in the light of what must, should, and could be known. The way in which time should be apportioned between theory and practice was dealt with in chapter 2. An understanding of the importance of each subject in the syllabus and the time required to teach it effectively is essential.

There is a tendency to allot time only on the basis of the *importance* of the subject, ignoring the time required to *teach* the subject adequately.

A simple example is given in figure 40, where each block represents a subject shown in its relative importance, assessed by its contribution to the object of the course, and therefore a course of say forty hours would be divided up as follows :

must know	A	10 hours,	B	7 hours,	C	6 hours
should know	D	5 hours,	E	4 hours,	F	3 hours
could know	G	2 hours,	H	2 hours,	I	1 hour

This, however, is not a sound method, because it is clearly better to decide how long is *really required to teach* A, B, C. Suppose the answer is thirty-two hours, then only eight hours are left to teach D, E and F. At this point the training supervisor should make the position clear to his seniors to whom he is responsible. He can teach either

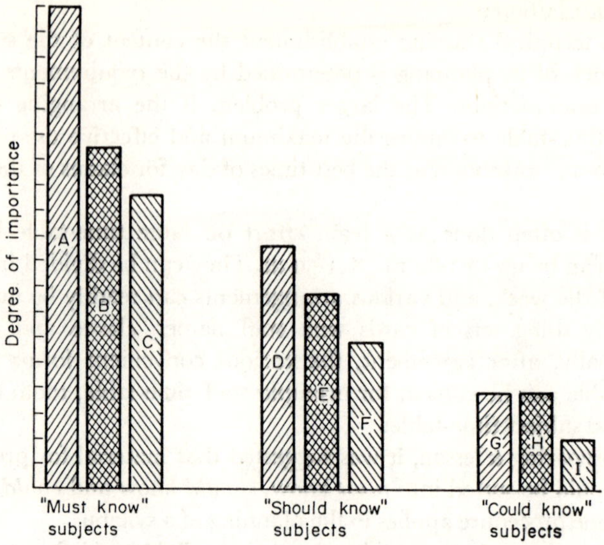

FIGURE 40 Syllabus plan related to subject importance

A, B, C and D and a part of E in the time allotted, or all subjects to achieve the full objects of the course, if given another twelve hours. Figures and arguments of this sort are cogent only if based on experience.

The first method of allocating time in rough proportion to the assessed importance of the subject, regardless of the time required to teach it, might produce a smattering of all the subjects with no real proficiency. The second and better method is based on an allocation of time necessary to learn effectively what is required, having regard to the importance of each subject.

Some technical courses which aim at keeping trainees up to date are more satisfactory if made short and given frequently, at say six-monthly intervals, than if long and given at say two-yearly intervals. This is specially true in electronics where developments take place rapidly.

An employee taken off a job and deputed for a course is apt to grow anxious if too long away from his normal work. A training course should not be put on when work is slack, as this creates the impression that training schemes are a means of maintaining a labour pool, from which men can be withdrawn as required. Such an impression will undermine confidence in the value of the training given.

It has been emphasised from time to time that the development of essential personal qualities in the trainees is an important part of any technical training programme and this aspect of training should not be lost sight of in planning a syllabus.

Some parts of the syllabus, such as general lectures, have very little character-training value, other parts, for example competitive practical tests in teams under elected leaders, have a high value. Parts of the syllabus involving team-work or perseverance bring out personal qualities without which the whole training programme can be a waste of time.

A valuable means of maintaining interest and class morale is to aim at getting trainees on to production work as soon as an acceptable standard has been reached. To know that all one's work is going to be scrapped irrespective of its quality, just because one is under training, is poor encouragement.

The planning of training courses is often made easier if a master syllabus be used as a guide. An example of a master syllabus for a record clerk to be given as an example (p. 147).

Refresher courses are important, for, although they devote a good deal of time to background subjects, they have as their object the maintenance of the employee's interest and loyalty, by keeping him up to date on the firm's aims and policy—especially on problems affecting management and employees.

It is important to have good relations and communication between a training section and production supervisors. Suspicion or bad relations lead to the new worker being informed by a supervisor: 'That's what they teach you in the school. Forget it. This is what we do here.'

Training as a whole

Technical training must be continuous and progressive, and should grow into a unified and coherent whole. This means that the subjects and subdivisions of each subject (which for convenience are often taught in isolation) should eventually be brought together in a well-devised exercise, or a comprehensive practical test job, to unify them.

As a homely example. When training a lorry-driver who will have to be responsible for some care and maintenance of his lorry, there are many aspects of the job to be taught. These should not be taught in complete isolation, but the inter-connections should be pointed out. To prepare the way to a unification of the subject-matter, the instructor should begin by taking the trainee out in a lorry to give him an insight into the job, and an understanding of what is required for skilled and intelligent driving. In this way there is formed early in training a conception of the whole body of knowledge and skills to be learnt. Sections such as the gears and clutch operation, the braking system, the electrical system, fault-finding, steering mechanism, and the Highway Code, even though necessarily taught in isolation, will not then appear to be so isolated.

A liberal amount of time should be devoted to trials which call for a full understanding of the whole job of driving and maintaining a vehicle.

This principle was advocated in chapter 2 when considering what sequence a lesson on a piece of mechanism should follow, and can be summarised as:

1. general function and purpose as a whole;
2. analysis into parts, and teaching each part separately;
3. final synthesis into the whole.

Standard lessons

It is possible for training to be standardised down to the smallest detail, and lectures or lessons prepared and duplicated by senior instructors and kept in cold storage for use when required. In effect this means that the training officer does the preparation for the class instructor. This may save time and may produce greater efficiency, but it has the serious disadvantage of robbing the junior instructor of any opportunity for initiative and destroys his enthusiasm for his task.

A standard lesson given without zeal or conviction is far less effective than instructor's own lesson given in his own way to suit a particular class. The lesson 'off the peg' should be regarded as useful in an emergency.

In planning a syllabus the instructor will find that, no matter how well he studies the importance of tempo, fatigue or inhibition, a number of other factors enter into the problem which have to be dealt with by compromise, goodwill and common sense. For instance, other classes may want to use the workshop, the film projector, or the drawing office at the same time. Again, some of the principles outlined may appear to be in conflict. It may be more important to the instructor to ignore the principle of fatigue in order to gain a valuable tempo. Variety in teaching may be achieved only by risking a little inhibition or confusion in some members of the class.

Compilation of master training syllabus for any grade of employee

Explanatory Note. The grade of *Record Clerk* has been used in the master syllabus on the next four pages, which is intended to be nothing more than an illustration of the *method* of compilation. The detailed content, and the number and duration of the courses required by record clerks in any particular firm, will depend on the type and complexity of the records kept—this illustration should not be regarded as a guide to these matters. In any firm, an induction course, followed by one or more job training courses, will be required by all new employees; but the need for promotion courses for a particular grade will depend upon whether the grade covers complex or responsible work. If it covers only simple routine work, employees suitable for promotion will not remain in that grade. Whether the work of the grade is simple or complex it is desirable to hold 'refresher' courses from time to time; the intervals between such courses can vary, and will depend upon the circumstances of the firm.

This syllabus is reproduced by kind permission of the Institute of Personnel Management.

Subjects	**Course 1** *Induction and preliminary training*	**Course 2** *Job training: simple records*	**Course 3** *Job training: fairly complex records*	**Course 4** *Complex records (promoted staff)*	**Total time required for basic training**	**'Refresher' courses** (1 *week, afternoons only*)
1. Welcome by company official	½ hour				½ hour	
2. Industrial background						
a. History of the industry	*a.* ½ hour				1½ hours	
b. Present position—products and markets; types of firm and their organisation	*b.* 1 hour					*b.* 1 hour
c. Current and future developments and problems						*c.* 1 hour
3. The company						
a. History: products and their uses; aims and policy of the firm	*a.* ½ hour					
b. Management—organisation of the firm; senior officials and their functions; work of main departments; tour of all departments	*b.* 3 hours				3½ hours	
c. New policies; new production programmes; new products; changes in organisation						*c.* 2½ hours

4. Legislation						
a. National insurance; income tax; workmen's compensation	*a.* ½ hour				½ hour	
b. Changes in relevant legislation						*b.* ½ hour
5. Health and safety						
a Firm's health services; accident prevention; fire precautions; procedure in case of illness or accident at work	*a.* 1 hour				1 hour	
b. Changes and developments as under *a*						*b.* 1 hour
6. Company personnel policy and procedures						
a. Conditions of employment and staff rules—hours of work; salaries, and method of payment; holidays and sickness absence; time-keeping and time recording; attendance and leave of absence; the 'staff code'	*a.* 2 hours	*a.* 1 hour	*a.* 1 hour	*a.* 1 hour		
b. Training and education schemes; promotion policy and opportunities	*b.* 1 hour	*b.* ½ hour	*b.* ½ hour	*b.* ½ hour	10 hours	
c Co-operation; staff council and committees; staff representation; function of the personnel department	*c.* 1 hour	*c.* ½ hour	*c.* ½ hour	*c.* ½ hour		
d. New policies and practices as under *a*, *b*, *c* and *d*						*d.* 3 hours

Subjects	**Course 1** *Induction and preliminary training*	**Course 2** *Job training: simple records*	**Course 3** *Job training: fairly complex records*	**Course 4** *Complex records (promoted staff)*	**Total time required for Basic Training**	**'Refresher' courses** (1 *week, afternoons only*)
7. Staff amenities and services						
a. Clockrooms and toilets; canteen facilities; rest and recreation rooms	*a.* ½ hour					
b. Pensions scheme; sick benefit club; benevolent fund; savings; sports and social club	*b.* 1 hour				1½ hours	
c. New developments as under *a* and *b*						*c.* 1 hour
8. Sources of information						
a. Methods of informing staff of new policies, rules and procedures	*a.* ½ hour					*a.* ½ hour
b. Departments and officials dealing with queries; policy and procedure covering complaints and grievances	*b.* ½ hour				1 hour	
						b. 1 hour

9. Background of the job						
a. Introduction to head of new department and supervisor; work of department and relationship to other departments; departmental layout and sequence of operations	*a.* 1 hour	*a.* 1 hour	*a.* 3 hours	*a.* 5 hours		
b. Departmental routines	*b.* ½ hour	*b.* ½ hour	*b.* 1 hour	*b.* 5 hours	17 hours	
c. Changes and developments in work of department; new departmental routines						*c.* 2½ hours
10. Specific job knowledge and skill						
A. Purpose and functions of records; types of records; firm's record systems	A. 3 hours		A. 2 hours	A. 10 hours		
B. Purpose of employee's specific job; duties and responsibilities; work methods and procedures; forms and symbols used (theory)	B. 7½ hours	B. 15 hours	B. 29 hours	B. 60 hours		
C. Job demonstration and practice under supervision	C. 15 hours	C. 22 hours	C. 44 hours	C. 82 hours	295½ hours	
D. Standards of work—accuracy, speed, neatness, economy; standards required and reasons for them; queries and errors	D. 1 hour	D. 1 hour	D. 2 hours	D. 2 hours		D. 1 hour
E. New types of work; changes and developments under A, B and D (theory and practice)						E. 2½ hours
Total time required for training	41½ hours (1 week)	41½ hours (1 week)	83 hours (2 weeks)	166 hours (4 weeks)	332 hours (8 weeks)	

A systems approach to the planning of a course

Now that we have considered the learning process a little more, and discussed the factors involved in planning a syllabus, we are now able to take a look at a suggested systems approach to the planning of a training course as shown on p. 153.

This diagram can be built up systematically on a felt-board or magnetic board, and can be displayed in conjunction with the 'training function' on p. 111.

Both will give rise to points for discussion.

There is a rough scientific method in this approach (figure 41).

1, 2 and parts of 3 represent the *analysis* stages of the system which identifies the problems and gets the facts.

Parts of 3, 4, 5, 6, 7 represent the way the problem is tackled, based on an *hypothesis* that it will work, and fits the facts.

8 and 9 represent the *results* and the *testing,* and the discovery of new facts and problems that can be fed back to the 3, 4, 5, 6, 7 for internal validation.

10 provides the all-important feedback to 1 and 2 from the job, and is the best evaluation of training success.

A systems approach to the planning of a course

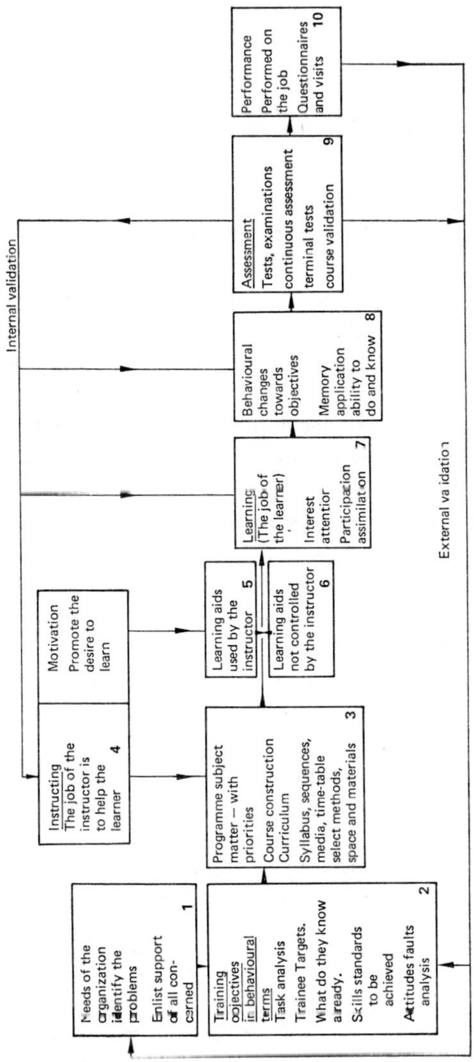

FIGURE 41

8

The Various Methods of Instruction

Instructors may be a little staggered by the ever-increasing number of innovations in methods of teaching and learning in both the hardware and software of educational technology, but we should not be depressed by the avalanche of published material on theories of learning, computerised learning, skills analysis, systems analysis in training programmes, assessment of needs, evaluation techniques etc., that has swept across the training scene in the past 15 years. In essence there has been developed a more systematic and scientific outlook on the learning process and the organisation of both the hardware and software of learning, taking into account the wide range of objectives in technical training and the needs of individual learners.

The instructor as we have seen in chapter 2 needs to have at his service a variety of instructional methods with which to enrich his instruction and the trainees' learning process.

In chapter 6 on assessment, the three main aspects of training: teaching, learning, assessing, and the close inter-relation between them were discussed.

In considering more fully the various methods of instruction it may be helpful to draw up lists such as showing the inter-related factors in some detail (see figure 52, p. 213).

1. the various *methods of instruction*;

2. the various *modes of learning*—these are inseparable from learning aids;

3. the various *means of assessment* which can be used to evaluate the instructor's success and the learners' progress.

It is not possible to draw a sharp line of demarcation between some methods of instruction and modes of learning. When a learning aid, such as a film of 20 minutes' duration, is used by an instructor and is of central importance in the piece of instruction, it could be considered as a method of instruction.

With the rapid expansion of the new subject, of educational technology, enthusiastic experts offer various pieces of hardware or new techniques as magic panacea for training ills, but the overall cost benefit of much that is done in the name of training is being increasingly brought into question.

Methods of measuring training effectiveness are admittedly not very satisfactory and the human relations problem remains.

The truth is that each method of instruction or mode of learning is effective within certain limits, and each has its own strength and weaknesses.

The lecture

A lecture has many limitations in usefulness but nevertheless has a place in technical training. For example, a lecture by the medical officer is usually the most satisfactory way of instructing a class on some aspect of industrial hygiene.

A 'pep' talk by the managing director or training supervisor could take the form of a lecture, as the objects of these talks would be to get men to think in a certain way; to motivate them, rather than to impart knowledge or skills.

A lecturer, if he is to achieve the object of inspiring men to be more industrious, loyal or punctual, must not only be a good lecturer, but an able and experienced leader.

Lectures have several advantages: one instructor can handle a large class of a hundred or more, he can cover a great deal of ground in his own way, no equipment need be used, and a lecture well prepared can be repeated without any great effort on the part of the lecturer by way of further preparation.

A Government Report on methods summarised the evidence in favour of lectures as follows.

'Immature students learn more readily by listening than by reading.

Lectures are especially valuable for introducing and opening up a subject and students can thus be led into subjects which would otherwise prove too daunting for them.

Where knowledge is advancing rapidly textbooks may not be available.

Lectures awaken a critical attitude in students.

Lectures can provide aesthetic pleasure.

Inspiring lectures can reach large numbers of students and are economical of the time of staff.'

To quote from *Communication and Learning*, by L. S. Powell.

'As with kissing, clinical tests can prove that lecturing is a "bad thing." Indeed the comparison can be taken further. Those who have never participated in a good lecture cannot know its power to inspire: its full impact can only be appreciated by consenting adults; its effectiveness is usually dependent in part upon effective visual aids. And, furthermore, despite all the protests, it is going to remain an important mode of communication for a considerable time to come.'

The main disadvantage is that knowledge gained by passive listening to a factual lecture, without some participation by the class, is not readily assimilated, and saturation sets in very quickly. The lecture offers no scope for class participation and is therefore of little use in the *training* of men in skills. The advantages benefit the instructor, the disadvantages penalise the class.

The old-fashioned form of lecture consisting of sixty minutes of monologue has no place in technical instruction. The class is passive, there is no variety, and boredom readily descends upon the class. Experiments on classes show that, after a period of sixty minutes of talking, only a fraction of the class can remember any more than could have been stated succinctly in ten minutes.

Although a lecture is of very limited effectiveness, it can be given its maximum value if the subject matter is interesting, relevant, well-prepared, arranged in logical sequence, and presented in an attractive manner, using audio-visual aids as much as possible. The principal weakness of the lecture method is that students become passive listeners. In all occupational training there must be active class participation, and as a general rule, at least half the time should be allocated to this.

The lecture can readily be transformed into a live experience both for the class and lecturer by reducing the time from sixty minutes to thirty minutes of well-planned talk, broken up into sections of ten minutes each by visual aids, periods of questioning, etc., to stave off saturation.

Technical training is mainly concerned with people *doing* things and therefore the most effective medium is that which presents the syllabus in the most practical and useful way.

The least effective medium is one which uses only words—without the support of things to see or do. Note, for example, the difference on a class of these two methods of approach to a lesson or a course on first aid.

1. 'I am going to lecture this afternoon on accidents, and what to do in emergencies. As you probably know, accidents often happen and people lose their heads and do the wrong thing. The first thing to do when you find yourself in the position of having to render first aid is to, etc. etc.'

After fifteen minutes of words the class will be losing interest.

2. 'Under this blanket I have a casualty from a rather serious accident (previously arrange a case of bleeding or a compound fracture). You are alone. What will you look for first and what action will you take? I am going to ask you some questions, then ask someone to deal with the situation. The others can later discuss what was done. I will then demonstrate a few points to illustrate routine procedure when accidents are first discovered.'

Here the class is active and interested : *sense appeal* is used.

In deciding how to transmit a lesson, the choice of method is often limited by consideration of size of class, time, and equipment available, but a keen instructor will overcome these limitations and insist on using the most effective medium for his purpose.

It has been emphasised that, in technical training, the factual lecture should be avoided as far as possible, but, if for reasons of time, and ease of preparation, the lecture is somewhat reluctantly decided upon, then make use of drama, surprise, variety, questions, visual aids, as discussed in chapters 2 and 3.

Skills cannot be taught by lecturing. Techniques can sometimes be introduced by a lecture, but time is often wasted by trying to describe in the classroom, processes and operations that can far more satisfactorily be demonstrated and practised on the job or in the workshop. Talk and chalk, gestures and diagrams, should be confined to subjects and aspects of training that cannot be taught by more practical, lively and active methods.

With a dull and difficult class of appentices in the lower grades of work the lecture calls forth a response mainly of passivity, but a bright well-motivated class will stand up to a lecture, and assimilate the factual material with reasonable relish.

Participative learning

Participation and involvement of the learner is an essential part of any learning method but this can take a great variety of forms. For example there can be some participation in a teacher-centred situation in which trainees are allowed or encouraged to ask questions of the instructor or tutor and thus make a little contribution and participate.

This, however, is nothing much more than a lecture method slightly modified to help the learners a little more, and to introduce a little feedback to help the lecturer—if he is able to make use of questions as helpful feedback.

There is also a much used method of old-fashioned apprenticeship training which is in a sense participative learning, and which is certainly not teacher centred. It is more generally known as 'picking the job up'. This is often a true description, as bits and pieces of the job are picked up by the apprentice in his long periods of waiting around. He can also pick up attitudes and ways of filling in time which are of negative value.

There is a point of view arising from modern educational theories that trainees should themselves decide what and how they want to learn. The teacher becomes a part of a 'learning resources centre' containing all appropriate learning aids which can be used as required or as felt to be appropriate by the learners.

This approach to learning gives rise to 'learner-centred learning', but there are dangers in this departure from traditional practices, as many trainees have never been encouraged at school to learn in this situation, and therefore expect to be taught and helped to learn by more authoritarian methods.

Punch graphically makes this point in a drawing of a small girl in a progressive education establishment, almost in tears demanding of an obviously kindly modern teacher: 'Must we always do what we want to do?'

How much telling, teaching, showing, demonstrating, participation, activity, doing, freedom to learn by mistakes, freedom to discover, in any particular class, lesson or course can be decided only by experience and taking into account all the factors, such as: the objectives, whether skills, knowledge, attitudes; the size of the class; the ages; the level of the work; the personal differences between the individual; and the personality temperament of the teacher. See figure 52, p. 213, 'What is involved in teaching and learning'.

Mutual lectures

The class is given some time to prepare various subjects, either individually or in syndicates of three, and then someone is selected to give a short lecture, followed by others. This method needs careful handling and supervision. These can be used only when the class-consists of men of fairly mature experience or advanced ability.

Advantages

The class activity in preparing and taking part in actual lecturing ensures good instruction and assimilation. Character qualities—such as leadership initiative—are given some scope for development.

Disadvantages

The disadvantages of the method are that much more time is required than for a straight lecture. If a large class is split up, extra instructors will be needed. If a student does not make a success of his lecture, the ground will have to be covered again by the instructor.

The advantages benefit the student—the disadvantages affect the instructor.

Lesson

This is a suitable name for the normal type of class teaching in classes of ten to twenty men lasting from thirty to fifty minutes, and making use of explanation, demonstration and practice, in varying proportions to suit the teaching of knowledge, skills or attitudes. It should be characterised by close co-operation between teacher and taught, with each man being given individual attention from time to time.

As we have seen in chapters 2 and 3, the success or otherwise of a lesson depends on preparation and the qualities of the instructor.

Project method

This method of teaching is one of the most effective ways of arousing interest. It is a reaction against the rather rigid and artificial divisions of knowledge and shows that a practical job or problem in real life demands a blending of several subjects. Building a canoe or a small sailing boat is a common project undertaken by one, two or more boys, involving the study of drawings, careful measurement, the use of tools and the assessment of qualities of timber, screws, nails, canvas, etc. Projects should be a creative and useful experience for the students and a good instructor will be able to devise them either for individuals or as a joint effort by a group.

Many trainees will have had some experience of project work in their schools, as this method of teaching is being encouraged more and more.

A training project should be so conceived that the trainee feels the responsibility for the planning and organising of the work.

He is expected to discover the facts and the relationships and the snags. He should have access to adequate information sources. The project should make a particular contribution to the class, the country or organisation in the form of a report or a piece of equipment.

Projects can exemplify the principles already discussed of feedback at regular stages and continuous assessment.

Projects can be very rewarding in terms of achieving behavioural objectives, as they demand knowledge, understanding, discussion and co-operation, analysis, synthesis, experiment and evaluation.

Project work has a strong element of discovery training in it, and learning by discovery is one of the most effective and valuable ways of motivating trainees who will enjoy finding things out for themselves. The method is essentially one which develops a trainee-centred learning situation.

Assignment and discussion

A classroom has been likened to a market of ideas, where experience and ideas are actively exchanged. The teacher should not monopolise the market or devalue the contributions of others. By the method of assignment and discussion, ideas are collected and pooled by teacher and students. A task or a problem is suggested by the instructor or by a trainee, which is a challenge to the trainee and calls for effort and initiative. The work to be carried out may involve experiments, handling samples, making things, sketches, diagrams, observation of films, use of the library, etc. After adequate time a discussion is held which, if skilfully handled, can be of great and lasting value. The method ensures active participation of the class and encourages initiative and the commendable habit of finding things out for oneself.

Discussion training is a useful method of motivating trainees, and is particularly valuable in changing attitudes through participation in group discussion.

Group discussion has been formerly the particular province of management, but discussions can enrich lecture and lesson material at all levels.

Group discussions

This is a useful method of instruction and of stimulating thought and interest in a subject or topic, and is generally conducted in small groups under a leader or chairman. Discussion groups contribute much toward character-development and training for posts of responsibility in dealing with people. In *Training within Industry for Supervisors* the discussion group technique is used extensively.

1. *Object.* To provide an opportunity for men under training to discuss, in a free and informal atmosphere, the various problems of the work and future. By expressing their opinions, the men are trained to think for themselves, and under proper guidance from the group leaders, the discussion group may help to achieve the following:

(*a*) individuals will be forced to face facts, and to give weight to the opinions and experiences of others;

(*b*) opinions will become more responsible, criticism will become more constructive and less destructive and judgments will be more objective;

(*c*) industrial problems will be viewed from a national rather than from an individual angle, and a sense of civic responsibility will be developed.

2. *Aims of the group leader or chairman.* These are:

(*a*) to arouse the group's interest, to keep the discussion general, balanced and relevant;

(*b*) to stimulate thought and discussion on the topic;

(*c*) to give the group information when necessary.

The following is a suggested general pattern that might be used:

1. introductory talk by the leader which conveys the purpose of the discussion and the possible lines it might take;

II. discussion by the group: *a.* pooling of facts known by the group; *b.* opinions; *c.* summing up by the leader.

A discussion is best run as informally organised conversation, between members of a small group, not more than twenty-five. It should not be used, or be regarded as a formal debate, and is not suitable for the communication of factual information. Those interested will inevitably seek information, and a great deal of information will be contributed by the group, once interest is roused. The aim should be, therefore, to achieve maximum collaboration in the group.

3. *Hints for the chairman or leader.* His job is to occupy the chair, not to monopolise the floor. He should see that the class is comfortable. He should state the subject, its various phases, and decide in what order those phases will be taken. Then he should put a question to a member of the group, to begin the discussion.

Thereafter, he should keep the discussion going by:

1. asking for further answers to his question;
2. asking opinions on the first answer he receives;
3. eliciting new questions;
4. posing new questions or problems.

He should see that:

(*a*) over-enthusiastic talkers do not predominate;
(*b*) over-timid talkers do not remain silent;
(*c*) all points of view are discussed—if necessary soliciting propositions opposing the dominating view;
(*d*) the discussion does not wander from the main subject, or become trivial.

He must help the class to see the implications of any opinions expressed. He should endeavour never to put forward his own views.

It is not necessary, though it is helpful, for the leader to have detailed knowledge of his subject, but it is *vital* for him to think the subject over to split it into convenient sections, and to anticipate, generally, the opinions that may be expressed and to realise their implications.

4. *Preparation.* Every discussion requires preparation by the group leader, as the success of it will depend on how the subject is initially presented by him and how it is controlled.

a. Subject for discussion. This should be a controversial problem of human relations or concerned with social or industrial matters. Articles in newspapers and magazines will serve as suitable bases for discussion.

b. Selection. This is a matter of some difficulty and the selection of material must be made with care. Having read the source of information, the group leader should then extract the most important points. For an ordinary discussion, it is rarely necessary to have more than three or four points, though it is as well to have one or two in reserve in case interest flags in the ones originally chosen. These points should then be arranged in a natural sequence, so that the discussion on one point will lead automatically to the other. It is not essential to assimilate a mass of facts around each point, for it is more important to collect thoughts around the subject and decide on an attractive and interesting method of presenting each one.

5. *Hints on technique.* There is no standard method of running a discussion group, for the method used will be determined partly by the instrinsic difficulty of the subject, partly by the background knowledge and intelligence of the group and partly also by the skill and experience of the group leader. The latter must experiment to get the best out of himself and out of the group. The following factors should be borne in mind.

i. The leader's personal opinions do not necessarily matter as the whole object is to encourage the group to express theirs. If, therefore, a factual talk is unavoidable, it should be as short as possible.

ii. A discussion group is not a lecture nor a debate, nor an informal conversation, though certain aspects of all three can enter into the proceedings.

Seating arrangements can greatly influence the style and nature of a discussion or study group. The common pattern with the instructor in front on a platform and trainees arranged in regular rows makes it clear that the instructor or tutor is the boss and the source of authority. Communication is easily made between the leader and the others, but not so easily made between individuals of the group and so does not encourage talking, communication, and participation between members, in a learning situation.

It is much more effective for the group to sit more informally, so arrange chairs in a circle, semi-circle, a square or a U shape so that everyone can communicate with everyone else. Move around if necessary to occupy the best position for the style of the instruction whether for a lecture, discussion, watching television, film or slides. Make sure that everyone can see and hear what he is expected to see and hear.

Use people's names frequently—scribble a quick seating plan to help you know who people are.

Decide possibly after discussing with a few trainees how you should address people—'Smith', 'Mr. Smith' or 'Joe'.

This depends on ages, how well people know each other and you.

As stated above, there is no standard method, but the following may provide some guide lines.

> Having given the group a general idea of the topic for discussion, the leader passes the subject to the group by means of a question. These opening questions should be expressed in clear and simple terms and should be addressed to the men who can deal with them. This starts the ball rolling, and when he feels that sufficient attention has been paid to this particular point, he will pass on to the other points in a natural way, as opportunity offers. At the end the leader should summarise what has been said so that members can go away with a clear idea of the arguments brought up on every side. The purpose is not necessarily to arrive at any definite decision. Alternatively, if the leader is certain that no discussion is possible without a preliminary talk by him in order to give the group the necessary background information, then his talk should be as factual as possible. But, even during this talk, the leader should do his utmost to get the group to co-operate, and he should try to build his talk on fragments of information and personal experiences, obtained from the group. The background knowledge having been imparted, the leader should then state what questions for discussion arise from his talk and he will then start the discussion. A good group leader will never say, 'That finishes the talk, now we will start the discussion.' The discussion should follow and spring from the talk in a natural manner. A summary of arguments at the end is again recommended.

The group leader's job is to stimulate the discussion, to see that each member has an opportunity of speaking, and to encourage all to co-operate. Tact and sympathy are essential if the silent members in particular are to voice their opinions.

The leader should generally avoid answering questions put to him. The questions should be passed to other members of the group for answering, for that is the only way to achieve maximum co-operation. By continually answering questions the leader separates himself from the group and the thread of the discussion becomes broken.

The leader should always endeavour to keep the discussion within the limits originally laid down by him. There should be no rigidity here, for often revelant points are brought up by members of the group, which have more bearing on the subject than the points originally laid down by the leader. It is essential, however, to keep to the main subject under discussion.

Do not allow cross-talk amongst individual members of the group. All remarks should preferably be addressed to the leader or chairman.

The leader should use tact in discouraging voluble members from monopolising the discussion. This should be done with great care, as the talkative member is always useful in opening or reviving the discussion when it starts to flag.

The successful discussion is one which is carried on as a topic of conversation in the canteen afterwards.

Deal tactfully with the following

1. *The man with a mission.* He is much more concerned in inducing the other members of the group to agree with him than to convince them that what he says is sound. He usually employs exaggerated modes of expression and his frequent use of emotional language may serve to impress some of the people in the group.

2. *The 'Yes' or 'No' Man.* It is a grave mistake in some circumstances to try to draw a sharp line of demarcation. This is what logicians call 'the fallacy of either black or white'. Truth has many sides and in many circumstances it would be impossible to give a direct and definite 'Yes' or 'No'. Could one, for example, draw a sharp line of demarcation between the sane and the insane or between the intelligent and the unintelligent.

3. *The man with an isolated example.* A man is often tempted to generalise from one particular case, and you will generally find that it is not something that happened to him personally—it is usually his uncle's friend's brother-in-law, etc. When someone brings up a selected instance, you should endeavour to find out whether it is a

representative and fair case by asking the other members of the group.

4. *The man who lays down the law.* When you hear someone starting an argument with such phrases as: 'It is indisputably true that——', 'Everyone knows that——' or 'No reasonable person can doubt that——', then you can suspect that this is a substitute for any convincing argument or demonstration.

5. *The man with favourite clichés*—or a newly acquired piece of jargon which happens to be the 'in-thing' at the moment, but only among a select few—excluding the group. As an example a cliché that frequently recurs in speeches and the press is 'vested interest'. The following lines by Sir Alan Herbert* deal with this particular one:

To a Parrot

A 'vested interest' is—what?
Show me an interest that is not.
If I save sixpence in a chest,
That is a vested interest:
You have, in your beloved wife,
A vested interest for life.
George has another in his cat—
But we are none the worse for that.
I beg you, do not use again
Expressions you cannot explain.

6. *The man who counters an argument by misquoting it.* It is an old trick to misquote or to extend immoderately an argument which it is intended to counteract. The extension often destroys the argument and we generally find that this type of man will also add false imputations.

Syndicate discussions

The main features of the syndicate method of teaching and learning are as follows.

A group dealing with a subject which has several important

*Quoted by kind permission of the Author and of Messrs. Methuen & Co. Ltd., from *Light the Lights*.

aspects is divided into four or five syndicates of five or six trainees and each is given three or four questions and specific references. Members of each syndicate distribute the tasks and discuss views and ideas, and give a report, or are primed to speak about one of the particular aspects of the subject. This method requires tact and initial preparation and guidance but it is motivating, produces lively discussion and encourages initiative and participative learning.

Playlets or role playing

A little play-acting has a great appeal, and the short sketch or playlet of even only a few minutes' duration, designed to instruct the audience, is a refreshing relief from the lecture, and a valuable part of the instructor's stock-in trade.

Examples have already been suggested earlier. Briefly, the *advantages* of the method are these.

Interest is easily roused. It teaches visually and aurally and is useful in giving a background idea in teaching techniques which have often to be acquired by imitating the example of experienced men : for example, in demonstrating the correct bearing and attitude of a foreman or supervisor in dealing with various man-management problems. The film 'Introducing a new worker' teaches several man–management techniques by simple and effective dramatisation of various situations.

Playlets can be performed before large classes. They create a pleasant atmosphere in training which sweetens a tired or 'browned-off' class. Properly carried out, they make for good friendly relations between teachers and taught as the class see the instructor in a new light and see a little more of his human side.

Role acting and case methods are valuable methods for developing supervisory skills, such as decision making and problem solving. Although of limited use in craft training, the methods are applicable to instructor training, where the development of attitudes are of great importance.

The *disadvantages* are in the time and effort required for production, and, unless care be taken, the dramatic or humorous effects tend to dominate the effort, and may even obscure the instructional points or the main object'.

The wrong-way demonstration put on as a playlet can readily

develop into a piece of slapstick comedy, and then, although it provides good fun, it teaches very little (see p. 42). The demonstration is dealt with on p. 171.

The playlet is clearly more suited to teaching attitudes than skills or knowledge.

Playlets on problems in personal relations and the techniques of man–management can be effectively followed by a discussion. Play-acting is sometimes called, perhaps more appropriately, 'role playing': for example, a foreman dealing with the aggrieved workman. This can be put on either as a spontaneous situation in the form of a test, with no preparation, or it can be a rehearsed situation and put on as a polished performance to show how the situation might be effectively handled.

There are opportunities of teaching ordinary lessons and some factual material by this medium. Let someone represent the trainee on the stage. He could act the part of a visitor to the works, or a new managing director going round asking questions of the foreman and some workmen, who explain correctly and skilfully various intricate points connected with their work. The B.B.C. makes excellent use of such little scenes when explaining technical matters to its listeners or viewers. The inquiring visitor or important person going round the job must be very inquisitive, somewhat ignorant, and just a little stupid so that the lower members of the class can follow, and all the questions they want to ask are raised.

In a playlet, the instructor or someone should be in control on the stage or set, who can maintain some contact with the audience, and hammer home a point, by a summary, or by breaking in on the audience, or by rounding on them in the music-hall style. The best effect of a playlet is achieved if it has incidental surprises and a climax.

Good casting is most important and it is often necessary to prepare a script to suit the personalities available, but remember over-acting is more common, and even worse than playing woodenly.

The producer should ensure that players speak slowly and distinctly, and act naturally. The playlet should be a very simple affair with perhaps a few makeshift props and a hundred square feet or so of free space.

The script can be in note form and players may, within limits, use their own words.

Tutorial periods

These are useful in that they encourage students to learn in their own way, with the instructor available for consultation, discussion and advice.

In learning by rote, for example, sets of figures or sizes, the student must make his own special effort by 'sweat and test', as was mentioned on p. 72. Opportunities for private study should be provided in any course as a part of the process of assimilation.

Brains trusts

These can sometimes be arranged with good effect in order to provide background knowledge or a general survey of a subject, and can be designed to replace the standard general-interest lecture. They can be used to give junior men an insight into some of the scientific and research aspects of the firm. Brains trusts must be first class, and the personalities involved must be experts, for example the scientific research staff might stand up to questions on the use of radio-active tracer elements in industry. The development engineers might deal with questions of future trends in design. All kinds of technical and scientific problems could be aired and referred to experts, either from within the firm or from specialists brought in from outside.

Simulators

For technical trainees in certain fields, the simulator is one of the most important training aids. It is designed to provide realistic practice for trainees in both knowledge and skills and they simulate conditions with life-like experiences or danger, and they can give an immediate feedback of results.

There is nothing new about simulation. The dressmaker's dummy is a convenient simulator. In sport punch bags are used by boxers, fielding machines to train cricketers, and scrummage machines for rugby packs.

Training in the complex skills of modern technology has called for specially realistic simulators. For example, simulated cockpits with all the controls producing realistic movements have been specially valuable for the training of aircraft pilots, and simulators have been used for operator training in highly complex and automated industries, in which inter-dependent sensitive controls have to be operated with great skill and care.

Such aids enable trainees to practice without the constant anxiety that they will make mistakes that will endanger themselves or the equipment. Modern electronic controls and feedback devices make simulators a training method that should be greatly extended in training establishments. There is plenty of scope for the instructor's ingenuity and inventiveness in designing simulators for their particular training needs.

Teacher-teams

It is well known in industry and in other walks of life that some highly paid jobs are often done by workers who spend only a part of their time exercising their special skill or knowledge. For the rest they are engaged on routine or semi-skilled work that could equally well be done by a less qualified assistant.

Consistent with the principle that manpower should be used efficiently and that all workers should be fully employed on tasks best suited to their abilities and experience, 'teacher-teams' are now being used advantageously where large groups have to be trained. Each 'team' contains well-qualified experienced teachers, some able teachers, some less skilled and, if need be, some partially trained instructors. The training requirements are analysed and divided into parts that will fully occupy each category of teacher, enabling them to work at their own level and as a team.

Team teaching requires a co-operative group of teachers who can get along well together and who can accept the limitations the method requires in their autonomy. A trainee may not appreciate being taught by a team of people rather than one person to whom he can go with confidence when he needs special help.

The method also requires organisation but it should be adaptable to changes. Here again team teaching should not be regarded as yet another revolutionary innovation in educational technology, as any well-run training organisation will have a staff working as a team, with meetings and discussions to decide who does what in the best interests of the trainees and the organisation.

Team teaching should be an object lesson and a source of encouragement for teachers who find themselves working in isolation, and constrained to build a little empire even to the extent of being suspicious of anyone appearing to encroach on their academic or technical territory.

Team teaching can flourish only when all engaged in teaching give top priority to the basic aim of their job, which is to help trainees to learn. This aim should transcend personal ambitions to build up a section or a department.

The senior member of staff or head of a department would lead the team which accepts responsibility as a team and shares equipment and experience in friendly co-operation.

Module training

In many crafts or operations there are several clearly distinguishable skills or small groups of skills that can be dealt with for training purposes as a separate unit or module in a job. Each module has a distinct training element, including practice and experience, and possibly a further education element. A recognised level of proficiency or a qualification can be achieved through the completion of a series of modules. Additional modules can be completed for a higher-level qualification. The duration of a module depends of course on its skill and knowledge content.

The module system is designed to provide flexibility in reaching definite national standards. It takes into account the wide variation in the needs of individuals and of the industry. It makes use of all modern analytical methods for acquiring knowledge and skills, to the qualifying standards in speed and quality. The module system overcomes several of the disadvantages of the old apprenticeship system, such as its inflexibility over ages, and time of training.

Demonstration

This plays a prominent part in technical training. It is the link between explanation and practice, and makes use of man's natural inclination to imitate. The interest of even large classes can be roused and held by a good demonstration.

There are some disadvantages:

A good demonstration requires a great deal of time and trouble for proper preparation.

There is always a danger that a demonstration will display several features simultaneously, and a person can watch only one thing at a time.

Some demonstrations suffer from a too rapid sequence of events, so that something new happens before the preceding point is assimilated.

A demonstration of a skill for imitation by the trainee should not last too long, as the only way to learn a skill is by doing—not by seeing.

A demonstration that has for its object motivation—for example watching the works expert in armature winding or foundry work —need have no such short time limit as the object here is to show an expert at work, and that is like looking at a work of art, a pleasure in itself.

A trainee is not concerned with a demonstration as a demonstration, but normally focuses his attention on the demonstrator, and puts himself in his place and prepares to *imitate* him. This is best done if the skill is demonstrated slowly and in easily assimilated steps.

A demonstration is in danger of showing too much, going too fast and continuing too long. It must work perfectly, for anything which fails in a demonstration seriously detracts from it.

'Wrong-way' demonstrations should preferably be a representation of a true-to-life effort rather than a caricature. As previously mentioned, the 'Crazy Gang' version of a wrong-way demonstration can be uproariously funny but teaches practically nothing.

The most effective demonstration of a 'wrong way' is to show exactly how it is wrongly done, so that when the mistakes are pointed out and discussed the class feel that they have been saved from that error. This may require a 'faults analysis'.

Demonstrations to large classes may profit from the judicious use of an amplifier and loud-speaker. This ensures that everyone can hear, and the big voice speaks with some sort of other-worldly authority, quite unconnected with that of any instructor.

Note. Make sure that the loud-speaker is as far away from the microphone as possible to avoid the feedback noise so commonly heard when amplifying devices are badly installed.

Instructional visits

These need proper preparation and 'follow-up'. Approach the senior official of the place to be visited and fix the time, place, and the conditions of the visit in writing, stating the numbers concerned and the general object of the visit. Thank him after the visit. Prepare the party for the visit with an explanation of what they are to see, giving some relevant facts and figures. It is often useful to give to various members a few special points to study, and after the visit get them to give the class their findings which can then be discussed. This rouses interest and ensures that questions will be asked about the more important features. Such a visit can be part of a project.

Aids as methods

When an aid to learning is of central importance in a piece of instruction, it can be considered as a method of instruction, and the following learning aids fall into this category.

Film

This is such an important medium of instruction that a separate chapter, chapter 9, is devoted to it.

Recorded sound

Although considered here as a method of teaching, recorded sound is more likely to be used as an aid to instruction in a lesson or demonstration, as rarely could a whole period be devoted to listening to a recording. Tape recorders are now widely used in teaching languages. For this they are incorporated in 'language laboratories'. Learners work to a programmed lesson, and their efforts are recorded and played back, and faults eradicated. Modern instruments using magnetic tapes make recording simple, cheap, and an effective medium of instruction for certain types of training involving a technique, such as using the telephone, or the recognition of sounds. Examples are the sound of machinery under various conditions and the recognition of faults. Recordings of lessons, discussions, meetings, interviews of all kinds, man-management techniques (such as issuing instructions or handling men under various circumstances) can be put to good use in training instructors, supervisors or foremen. The recordings can be used to demonstrate good points in technique or to eradicate faults. They can also be incorporated into playlets with good effect.

Recorded instructions

The use of recorded instructions for the guidance of trainees learning craft skills is a recent development in training aids. It has the advantage that the trainee, using small earplugs, can keep his eyes and attention on the job instead of having to cope with charts, cards or instruction books, but the script must be carefully and expertly prepared. Further advantages are that its pace can be controlled by the trainee while its impersonality eliminates the uneasiness that often comes from being closely supervised.

It is salutary, and quite often humbling, for an instructor or foreman to hear a recording of his own voice. We all seem to possess vowel sounds and affectations in our speech that we little suspect until we hear ourselves as others hear us.

Guard against using a sound-recording as a substitute for a more realistic and practical demonstration by the instructor. A recording can be of value and convenience to the instructor by bringing to the classroom at the required moment sounds, notes, noises, speech or music.

Programmed instruction using teaching machines

This is a method of instruction of growing importance. Programmed learning is not new, but in recent years many devices for presenting programmes have been produced.

There is a danger that the fascination of the hardware and the commercial exploitation of gadgetry may obscure the simple basic functions and principles of programmed learning.

One authority in this field has commented, 'programmed instruction seems to be one of those activities that people are willing to play around with for a long time but never come to grips with. In some cases it may prove to be an eternal experiment.'

In a sense all good instruction is programmed in so far as it is planned, analysed and broken down into a sequence of small steps. The main advances have been made possible by presenting programmes to individual people by trainee-operated teaching machines (See p. 208).

These devices are unlikely to take the place of teachers, but they can lighten some parts of their work and may replace certain kinds of textbooks. Innovations in educational technology do not usher in conditions calling for less responsibility or effort from the teacher, but rather for more understanding of the learning process and the way humans behave.

Programmed learning as provided through teaching machines is successful because it is based on the principles of good teaching considered in chapters 2–5. The subject-matter is presented to the student broken down into suitably small steps or 'frames', each one generally in the form of a question requiring a response. As soon as a response is made, its correctness or otherwise is communicated to the student; each student can thus go at his own pace, and the teacher is free to give individual attention where necessary.

Many adult trainees are rather reluctant to learn in competition with others in a group. Programmed learning does not expose a person's weaknesses to others, but he has the satisfaction of enjoying successes and progress.

Programmed learning exploits to the full the process of step by step reinforcement.

It has many of the advantages of the tutorial method of teaching and learning, as it uses questions and answers, leading on to further questions until correct concepts are formed by the learner, understanding is reinforced, confidence is gained through success, and a desire for more learning is promoted.

The programme must have a clear objective stated in behavioural terms and great care must be taken in breaking it down into a planned sequence, and in formulating the questions so that they are stimulating but not so hard as to be discouraging. It is planned for a tutorial type of lesson to one or two students and ensures maximum student participation throughout. An average student should be able to get about 75 per cent of the response right; much less than this leads to loss of motivation. It has been observed that children work well with about 90 per cent success. The student usually finds the machine a challenge and tries not to be beaten by it; also it lets him know immediately just how he is getting on at all stages, and this we have seen is in itself a motivating factor. Programmed learning is quick 'feedback' learning.

Programmes used in teaching machines can be either *linear* or *branching*.

With the *linear* programme, the material to be taught is broken down into very small steps designed to encourage the correct response to the questions. It is specially valuable in teaching facts, definitions or the basic knowledge necessary for learning skills.

The *branching* programme is somewhat similar to the multiple-choice type of testing, and presents the material to be taught in larger units than is suitable in a linear programme. Explanations and further guidance are provided for both right and wrong answers, and students are encouraged to reach the right answers by proper reasoning through understanding.

The material given in the multiple-choice test in Appendix I could be used as the basis of a branching programme for a teaching machine which had four buttons A, B, C and D. If the wrong button is pressed, a note is given on why the answer is wrong with guidance towards the correct one.

There is no need for any complicated hardware or even electrical circuitry for this kind of programme to be presented to a class, as a simple 5 cm cube can be effectively used for this as explained on p. 208.

It would be a valuable exercise for technical teachers to prepare a lesson from time to time as if it had to be presented through a teaching machine. This would ensure a clear objective, a careful selection of material and its arrangement in a logical sequence, small steps, and above all the development of the main points by questioning.

Programmed learning has its limitations. While it may be effective for helping people to learn facts and to understand principles, it does little to encourage original thought, initiative and imagination. The learner is rather lead by the nose through a programme along a narrow path towards a definite goal, and is offered no particular incentive to ask his own questions and to seek and discover the answers, for which a discovery method is specially suited.

An Association for Programmed Learning has been formed consisting of teachers representing a wide variety of educational interests/research/and workers in industry and in education.

A note on the various kinds of teaching machines is given at the end of chapter 10.

Fault-finding methods—cause and effect charts

Modern electronic equipment has become very complex so that

tracing faults and training men to do this can be a formidable task, particularly if attempted without a logical system of excluding possibilities step by step.

This is now done by the development of charts which simplify the task through functional diagrams and instructions that are specific, effective and simple to follow, and can be in the technician's own language. The analytical algorithmic approach reveals the inter-relationship of all elements of the system, aids understanding and minimises the need for extensive experience or theoretical know-ledge. The charts and flow diagrams represent the thinking of a highly skilled technician in isolating faults, and is programmed to get the job done, and to help the learning processes of the less skilled workers.

Techniques once learnt can be used on systems of different com-plexity with a minimum of additional training.

A note on methods for older trainees

In these days of rapid advances in technology it is likely that courses will have to be conducted for older workers to bring them up to date, or to take on new and perhaps more complex jobs.

The older trainee is naturally not very happy about being a trainee again and may react against being back in a class-room where they may retain memories of an oppressive authoritarian school atmos-phere, associated with being told what to do all day. He may feel that 'you can't teach an old dog new tricks'. This is not true, as it is found that older workers can learn as well as young ones under proper instruction. He may try to overcome his fears by working excessively hard or withdraw from participation and say, 'I'm nearly sixty so you can't expect much from me'.

Some older workers will resist having to give up old ideas and acquiring new ones, for example over going metric.

All people now, young and old, are exposed to instruction or teaching programmes on technical subjects on the television or radio, some follow systematic courses, and meet for discussion and re-inforcement.

The instructor may at times be called upon to discuss a film or T.V. programme with trainees, so keep up to date.

In older people short-term memory tends to decline, and people forget factual information. This is largely because of reduced attention or interest given to the new pieces of verbal information.

The older people grow the more difficult they find it to learn by being told things; when older trainees are allowed to learn through their own efforts and activity there is good progress. This applies in fact to all ages, and the lesson here for the instructor is to make all instruction at all stages active, leaving trainees to work things out for themselves under suitable and restrained guidance.

Active participation of trainees in a learning situation is always preferable to passively listening to someone talking.

Multi media

Individual instruction has been developed to a high degree in audio-visual teaching machines or multi media teaching machines and has been applied to a series of modules in instruction in auto mechanics.

The trainee is able to work his way through a course by using a series of booths in each of which is installed a cassette cine projector, a 35-mm slide projector and tape-recorder–playback machine, together with work bench for his actual work.

The instructor keeps in the background, but is ready to help on request.

9

The Use of the Film in Technical Training

The film is a valuable aid to technical training and there are many effective ways of using it. The *advantages* of the film over other media are briefly these :

1. it brings a variety of exciting situations of the outside world into the classroom;

2. the instructor's task is lightened by a film, but not eliminated;

3. expert treatment of a subject, and the results of much skill and time in planning, are made available to the class;

4. the performance runs reliably from beginning to end, if properly conducted;

5. the film is a means of introducing variety into the instruction;

6. large classes can be dealt with;

7. repetition can be easily effected as required;

8. the attention of the class can be focused on a small area of interest, by 'close-ups' and magnification;

9. all the tricks of emphasis known to film-makers, notably animation, can be used to make vivid impressions on the class;

10. it can survey a broad field of study in a short time and can be used to reveal and explain movements vital to the proper understanding of many subjects, through the use of slow motion, and time-lapse techniques.

11. information can be recorded and shown to a class that would be impossible or perhaps dangerous by any other means, as the camera can operate in places inaccessible to people. It can bring together in proper juxtaposition scenes and sequences that would require much time and effort if the class had to follow the route taken by the camera, for example different sections and processes in a large power station or works.

12. the film is ideally suited for motivating people or influencing

their outlook. The tremendous possibility of dramatisation and portraying problems in the field of human relations makes the film of exceptional value, for example in induction training, and in providing background information for a course.

The *disadvantages* are the following :

1. There is no opportunity during the showing of a film for active participation by the class.

2. Contact with the teacher is lost or weakened, and he cannot even adjust the pace of the film to the rate at which the class can learn, in fact the instructor during the presentation of a film hands his class over to it. There is thus a serious danger of saturation and poor assimilation in factual films.

3. The cinema is inseparably associated with relaxation and entertainment, so that students are tempted to regard a period in which films are shown as one calling for no effort. A film can impart only experience through the eyes and ears and is therefore of limited value in teaching manual skills.

These disadvantages can be largely offset by the instructor observing the following points.

1. Select the film to fit in with the syllabus. Use it as a learning aid at the right time with a specific objective, not as a means of filling in an odd period.

2. A film show requires preparation. See and study the film beforehand. Titles can be misleading. Divide it up into two or three sequences if necessary.

3. Introduce the film explaining what the film is about and how it fits into the course and what to look for. It may be useful to set some problems for discussion afterwards as a test, and to prepare a blackboard summary that can be shown after the film to confirm what has been learnt, and to be taken down as notes.

4. A silent film can sometimes be just as effective as a sound film particularly if the instructor himself supplies the commentary to suit his lesson and local conditions, for example where there are language difficulties. Many projectors are fitted with a socket for a microphone that can be brought into operation as required.

5. Most films on technical subjects can usefully be shown twice with questions and discussions in between. In teaching a skill, a film loop that repeats skilled operations as often as required is very effective.

6. Films that are designed to motivate a class, to boost morale or

to affect personal relationship should be shown as a whole and without interruption. Comment and discussion afterwards may, however, be useful at the instructor's discretion.

A training film must have a clearly defined objective

One of the hall-marks of a successful lesson, lecture, or demonstration is a *definite and limited objective* (chapter 1). The same is true of a training film which should neither attempt too much nor cater for too many kinds of audience. A statement of the object should include a definition of the audience for which the film is designed, for example 'to instruct British farmers to recognise and deal with the Colorado Beetle'.

The film made with a strictly limited objective—the 'single concept film'—is likely to achieve even more than its specified purpose, but a film that attempts to instruct a vague audience in an ill-defined skill or technique will fail altogether.

Two simple questions must be asked of an instructional film: 'What is its object? How successful is it in achieving this object?'

Frequently the title of a film will indicate the object of the film, for example 'Precise measurements for engineers'.

Some instructional films made by industrial companies suffer from the lack of a clearly defined object. The worst contain some attempt at giving instruction in a manual skill, mixed with a wealth of information and factual knowledge, and throughout the whole film there is a feeling that an attempt is being made to carry on some sort of sales propaganda. Such films seem to be more concerned with impressing and justifying themselves to the board of directors than with training workers.

Film makers should be guided by a clear and limited objective at all stages of production: at the stage of planning the scripts, and during consultations between subject specialists, educational specialists and film specialists.

It has been found that pre-production planning is not always an unqualified success when done in committees as this often leads to unsatisfactory compromises.

A training film generally falls into one of four types depending on whether its main objective is: 1. to give technical information (knowledge) for example, *force, mass and acceleration*; 2. to demonstrate a physical skill, for example *use of the wood chisel*; 3. to demonstrate a technique, for example *brazing*; and 4. to influence attitudes and ways of thinking or to boost morale, and convince the trainee that his work is worth while and part of a much bigger industrial effort.

The film and the associated media T.V. are especially suitable for instructors and for motivating people in matters where human relations and personnel problems arise, and topics under 3 and 4 lend themselves more readily and usefully to the medium of the film rather than topics under 1 and 2.

In a course for foremen and supervisors, a number of short films depicting problems in human relationships can very profitably be shown and then discussed.

How to obtain films

The sources of training and educational films are many, varied and constantly changing, and a list of specific sources quickly goes out of date. A short one is given at the end of the chapter, but it is advisable to write to the Foundation Film Library, the Educational Foundation for Visual Aids, The Central Office of Information, Central Film Library, British National Film Catalogue, Sound Services, and secure their latest catalogues, which cover most subjects and producers and, being frequently revised, will lead one to all but the most recent suitable films in any particular category. These catalogues will also enable the user to acquaint himself with the producers of those films most relevant to his special needs and hence to follow their productions with special attention.

The first consideration in selecting a film must be the answer to the question, 'How does the object of a particular film fit in with the object of the course, or the period of instruction?' Films are cheaper when bought or hired 'off the peg' than when 'made to measure'. Most film-users have to use the best films that are available for their purpose; they cannot indulge in the luxury of having a film made specially to meet their particular requirements.

Users of technical films will find it advantageous to maintain some simple assessment cards of technical films seen. A card-index will help future instructors in selecting films. A suggested form follows and is based on the analysis of a training film into its main characteristics as shown on p. 183 and 184.

An instructor should always see and study the film he intends to show before showing it to the class. Titles can be misleading, and careful preparation on the part of the instructor is necessary for a satisfactory film period. It should also be introduced before showing and, if appropriate, discussed after showing.

How to show a film

The technical teaching film should not render the instructor redundant; there is a good deal of preparation work to be done before a technical teaching film can be shown effectively, and showing a film can often follow with advantage the same steps as for a piece of instruction.

Films designed to motivate the class should be complete in themselves and should be shown through without interruption or comment except for a discussion at the end if this is likely to be of value.

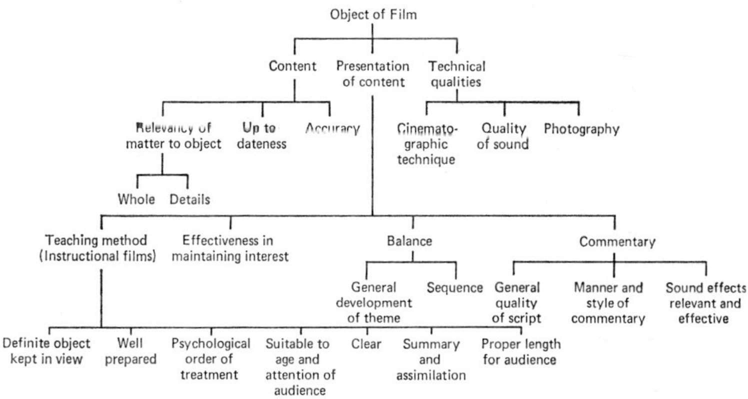

Film Appraisal

FIGURE 42

The following notes may help you to make full use of a film.

Preparation

The subject-matter (study film beforehand). If necessary, divide the film when selected into sequences of about ten minutes each, and select problems and questions to ask the class.

The minds of the class (lead up to film and introduce it at right stage in the syllabus). Give statement of the object of the class. Warn class of tests and questions to follow.

Materials and equipment (lighting, microphone, projector) so that no hitches occur. Electrical connections: is the flex adequate, and

SUGGESTED APPRAISAL FORM FOR SCIENTIFIC AND TECHNICAL FILMS

Title of film— *Running time*— *Colour or B and W*.—

Object of film including a definition of the audience or audiences for whom the film is intended—

Qualification of viewers (cross out whatever does not apply and add a note if necessary)—

Professionally qualified in subject-matter
Educationist
Film producer or specialist in cinematography
Typical of audience for which film is intended

	Yes	*Comment, if doubtful*	*No*
Film content			
Was the material well selected and relevant to the object of the film?			
Was it accurate and up to date?			
Presentation of content			
Was the presentation balanced with good sequence?			
Was interest well maintained?			
Was the script well written?			
Was commentary well spoken?			
Technical qualities			
Were the sound effects good?			
Were the photographic and cinematographic qualities good?			
General success of film			
Was the film successful in what it set out to do?			
Audience suitability			
Primary schools			
Secondary schools			
University and technical colleges			
Specialists in subject			
Industry—management			
technicians			
craftsmen			

Further comments if any

long enough for a safety lashing or clove hitch round the leg of the table or projection stand? Are the plugs and sockets of the right kind? Spare fuses, lamps? Can you focus on the screen? Is the picture of suitable size, i.e. is the projector lens of the right focal length for the room? In preparing a film for showing, thread it and run through to the title. Have the amplifier warmed up, so that title and sound come on together as the lights go out. Understanding with the projectionist, if far away use a coloured light code.

Presentation

For a factual instruction film this should ideally be done in ten-minute sequences with breaks for discussion to avoid saturation, then repeat (see chapter 7, p. 125).

Assimilation

This can be assisted by simple tests on the subject-matter of the film, giving back, question and answer. Summaries by the instructor. Note-taking.

Drive home each point by summaries, and 'part-way' summaries.

Do not use a film designed for a specific part of the syllabus, on a Friday afternoon when you cannot think of anything else to do!

General points about films

Makers often attempt to cram too much into a teaching film. Frequently, a film of twenty minutes' duration will attempt to teach far more than would be attempted in forty minutes in the classroom.

The silent technical film is sometimes to be preferred to the sound film, as the instructor, after proper preparation, can supply the commentary to suit his particular lesson and local conditions. Sound films can always be run as silent films if the instructor so requires. For large classes a microphone and amplifier is recommended for the instructor. One hundred words per minute seems to be the optimum speed for a commentary in a teaching film, and the commentary should preferably be a little in advance of the picture so that cue words should appear to trigger off the picture. Simple, direct speech ('hold the file horizontally') is better than indirect ('the file is held . . .').

There is no advantage to be gained from a musical background, but colour often has great advantages over monochrome, especially in animated diagrams, where it can increase clarity and vividness. Film devices for gaining attention which are not relevant to the subject-matter serve no useful purpose.

Some films should be shown twice for full effect, and, in order to ensure that this is done without technical inconvenience, the splicing of two copies of a film together has been recommended. Some films, especially those of the fast-moving kind packed with facts, are more useful for revision purposes than for initial instruction in a subject.

The efficacy of the film is greatly reduced by technical hitches, poor projection, bad acoustics, inadequately controlled lighting, lack of ventilation, and poor seating arrangements. Adequate preparation should eliminate these.

The instructor, or whoever is in charge of the screening, should be out of sight and will find it advantageous to have some simple means of communicating with the projectionist. One way out of many is to use three small coloured bulbs and some prearranged code of signals, to indicate; out lights—start film; stop film—on lights; increase sound volume; decrease sound volume; switch over from sound track to microphone; run same film through again; show the next film.

A film can be used effectively in job analysis, in time-and-motion studies, and in research on teaching techniques. It provides many opportunities for conducting investigations on audience reactions and has opened up many fruitful lines of inquiry.

The film should not be used just because a film exists on the subject being taught. Some films attempt to teach what can clearly be taught more successfully by other methods, for example by actual demonstration and practice on the real thing.

Single concept film loop

This is simply a device for projecting repeatedly a film which has had its end joined to its beginning, and by this means complicated physical skills or movements of mechanical parts can be isolated as a short sequence and repeated as often as required. Several loops can be used to ensure mastery of each operation in isolation, and then a complete film shown of the whole process. This is an excellent way of presenting such subjects as the working of the four-stroke engine, the gear-box or the differential gear of the car.

Film cassettes

A development which has made cine film much more available and convenient for classroom use is the 8 mm cassette film loop which can be projected in daylight through a translucent screen. The whole apparatus is light, portable and ready for instant use for demonstration, and all kinds of manual and athletic skills.

The film loop is a specially valuable aid as it can repeatedly show a particular skill—*as seen by the performer.*

Many demonstrations of a skill are seen by trainees at a spectator's angle which may give wrong impressions.

Cassettes are becoming available in a wide range of subjects, and they have many advantages over the 16 mm films in reels. Daylight operation by the teacher from the front of the class ensures continuous contact with the teacher. They can be changed in a few seconds, and as the film is in loop form, it can be shown several times, without interruption for rewinding and rethreading the machine. The equipment is much cheaper than the 16 mm equipment, and even taking films on an 8 mm camera is within the means of individual teachers interested in making instructional films to suit their particular needs.

The films may be silent or may carry sound. A sound film takes over from the teacher completely and there is much to be said in favour of the silent loop film used as aid to the teacher, who is in full control and can comment as much or as little as he feels appropriate. He can closely relate the film to his own teaching either past or future. He can stop or repeat the film at will, without undue dislocation or inconvenience.

Film flash

Occasions may arise in a course when a few minutes of film can be used with great effect to create an atmosphere, or to serve as an introduction to a lecture or a course, for example, on safety precautions.

Finally it must be borne in mind that the young trainee is a severe critic of films and television presentations and demands a high technical standard from any film if it is to claim his attention and approval.

Television

As a medium of instruction this is of growing importance, and is already widely used for demonstrating techniques to groups of specialists in widely separated areas, and for bringing into the shop or classroom outstanding technical events of the outside world, as they are taking place. Live television provides a strong 'here-and-now' interest to certain events, but, if simultaneity is of no importance, then it has no special advantages over film and video tape recordings. The importance of television as a mass medium in education is evidenced by the rapid growth in the past few years of educational broadcasts in all school subjects and the inauguration of the University of the air or the Open University, which is bringing to all who wish to have it a University standard of education in a wide range of subjects.

The Centre for Educational Television Overseas, Nuffield Lodge, Regent's Park, London, N.W.1, gives help to educational authorities overseas in preparing television programmes for their own countries. Educational programmes are now available to vast developing areas of the world, through the use of satellites orbiting the earth.

Closed-circuit television

The success of this teaching aid has been amply demonstrated in many technical colleges, and it has great possibilities for wider use, particularly in teaching skills and communicating ideas in situations that make direct observation difficult, dangerous or impossible. It has an advantage over the film in that conversation and communication between the teacher and the class is more easily accomplished.

Experiments in this medium indicate that its efficient and skilful use can help to overcome the shortage of teachers. The danger in using either television or film in a classroom, is not that it is likely to replace the teacher, but that it may tend to replace practical experience, and class participation. It provides however an excellent means of promoting class participation for teachers under training, by recording for discussion examples of micro-teaching by trainees (see p. 49).

Well-arranged T.V. and film demonstrations play an important part in instruction, but they should not be put on because they are easier to arrange than to set up a practical demonstration with opportunities for individual participation and practice. Simple demonstrations using magnets and iron filings for example, look very unreal on a television screen and are much better carried out in the classroom by the trainees themselves.

Some sources of training films

Central Film Library, Government Building, Bromyard Avenue, Acton, London W.3.

Central Film Library of Wales, 42 Park Place, Cardiff.

Foundation Film Library, Brooklands House, Weybridge, Surrey.

Industrial Society, Robert Hyde House, 48 Bryanston Square, London W.1.

Overseas Visual Aids Centre, 31 Tavistock Square, London W.C.1.

Scientific Film Association, 3 Belgrave Square, London S.W.1.

Scottish Central Film Library, 16–17 Woodside Terrace, Glasgow C.3.

Sound Services Film Library, Kingston Road, Merton Park, London, S.W.19.

10

Further Training Aids

Visual aids

In chapter 2 stress was laid on the value of using the sense of sight in instruction. Visual aids are essentially aids to understanding and learning and are not to be confused with aids to vision! Effective training is dependent upon good communication between teacher and taught, and this means that the teacher must be able to evoke, in the minds of his students, his own mental images and thought processes that he himself has established on the topic.

Words are a convenient and general means of communication, but they are not the only means. Alone they are frequently quite inadequate to convey an idea or a meaning, and even a simple sentence can communicate different meanings, depending on how it is said.

We have all experienced the difficulty of following verbal directions when we stop our car in a busy city to ask the way to a street a few miles away. Unfamiliar street names and verbal descriptions are often quite inadequate for communicating what is in the mind of our helpful direction giver, whereas a map or a sketch would have simplified and clarified his instructions.

The chart on p. 25 'How we learn through our senses' has a subheading 'A list of aids to training' because every one of the items listed can be regarded as an aid to training. Many instructors believe that the talking part of their instruction is of primary importance, and the most effective, and that the models, charts and demonstration equipment are mere aids in the process; but in some kinds of technical training it is the other way round, and the most effective and efficient means of communication is through the eyes. Visual aids attract attention, increase interest and introduce variety and change, but it is part of the instructor's art to choose his aids to suit his particular purpose. The working of a four-stroke petrol engine could be explained by using a real working engine, a sectionalised engine, a model, a film, a film-strip, an engineering drawing or blackboard sketches.

A good sectionalised engine would probably be the best choice, but this could be supported by other aids, for example a filmstrip and blackboard sketches. The real thing or an actual part for study is generally a better visual aid than a representation and should be used whenever possible. Models and three-dimensional representations are generally to be preferred to pictures and diagrams.

Here are a few points to be observed in selecting or preparing visual aids for class instruction. They should: (1) aid and not unduly dominate the instruction; (2) be simple with all non-essential details and lettering eliminated; (3) preferably be coloured, using two or three bold colours, provided the colouring separates parts and adds to clearness; (4) concentrate on one or two essentials or key parts at a time, and then build these parts into the whole; (5) be used to supplement information to be gained by handling the real thing, never as a substitute for it, and (6) should be shown to the class at the right moment, and then covered up or put away as soon as they have served their purpose. To leave visual aids to decorate the classroom is to invite inattention, so keep the front and side walls of the instruction-room free from any permanent posters, visual aids or pin-ups.

Posters and wall charts

These are intended to rouse the interest of trainees and may be advantageously displayed in workshops, on noticeboards, or at the foot of a stairway. Charts displayed in this way should be changed regularly, or interest will wane. A wall chart for permanent display should be of a different style from the classroom visual aid and may be an attempt to summarise a whole lesson or a course, and contain a great deal of detail and explanation. It should attract attention, be self-explanatory, be artistically produced, and encourage people to look at it, and to study it in odd moments. Charts for use in formal instruction, however, should be simple, clear, easily seen by the class and have one main point. Charts and diagrams used for instruction must be big enough to be seen clearly but not so big as to be unwieldy. Many firms and associations now produce excellent wall charts and visual aid material that can readily be obtained for teaching purposes.

The recognition by training officers of the importance of teaching aids in all their variety is one of the marks of the training revolution in industry that commenced about fifteen years ago. A list of publications and addresses useful for training officers is given at the end of this chapter.

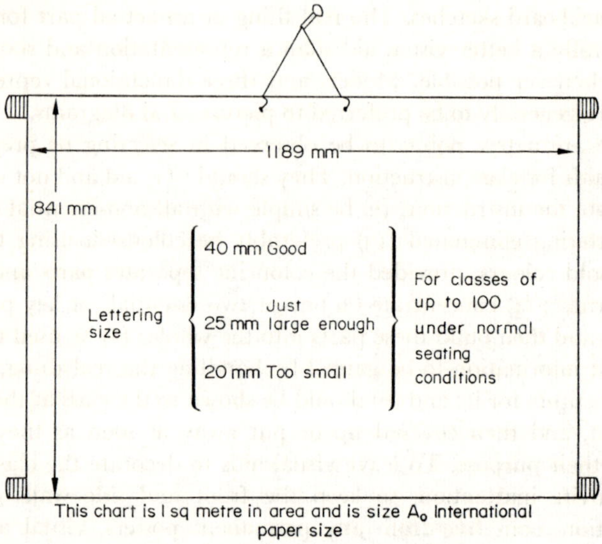

FIGURE 43 Wall chart: good standard size

Figure 43 is an example of a good standard size for a chart. Many firms supply charts on technical subjects, printed on paper of lightweight quality. Such charts when intended for constant use should be pasted on wall-boarding and varnished, or else mounted on linen and fitted with rollers. This is best done by a professional map or picture mounter but a simple expedient is to tack a lath to the top and bottom of the chart so that it can be rolled up, tied with a slip-knot, and let down instantly before the class when required (figure 44). The instructor should always test for visibility before his class assembles.

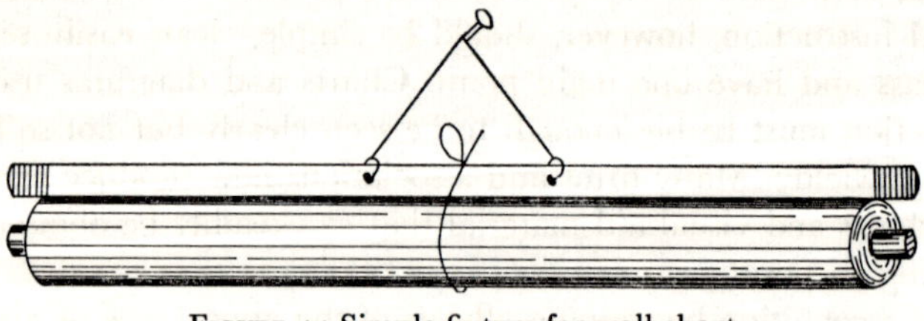

FIGURE 44 Simple fixture for wall chart

Charts are specially useful when shown in a sequence, to illustrate the systematic build-up of the subject or piece of mechanism. A series of charts, diagrams, graphs, etc., can be drawn on newsprint or other cheap paper using coloured felt pens, and then clamped together at the top. This is called a 'flap sequence'. The series can be conveniently hung over a blackboard (or the back of a chair placed on a table) and shown in correct sequence.

By interleaving the charts with sheets of blank paper, the instructor can cover up each of his charts at will without fuss or inconvenience as soon as it has served its purpose.

When charts or models have to be changed several times, the help of an assistant can greatly improve a lecture or lesson. The instructor should ensure that the attention of the class is focused on one thing at a time.

A quick and convenient way of fixing a diagram or chart to any surface is to use two lengths of Sellotape, about 10 cm long, not in the ordinary way, but twisted into a spiral with the sticky side outwards. This provides a short sticky roll which can be pressed hard between the chart and the other surface.

In conformity with the growing demand for metric standards, two useful sizes for charts for classroom use are the international sizes A0 (one square metre), 1,189 mm × 841 mm for large classes, and A1 (half a square metre), 841 mm × 594 mm for small classes.

Chalkboard

This is the oldest, and the cheapest of all training aids, and is still the most used teaching aid. Practically every technical subject requires a drawing to help in explaining it. In any technical discussion the participants rush either for a pencil and paper or to the blackboard. (In some progressive firms, every office has a chalkboard on the wall which forms an indispensable link in the communication process between the two or more participants in a technical discussion.) Usually white or yellow chalk is used on a black or green surface, which for durability can be covered with a ground glass sheet. Chalkboards can be primrose colour in which case dark blue chalks are used with advantage.

Blackboards, however, should be well maintained and should be black. They are often, through much use, grey and patchy. Indian ink will renew the surface of a worn board. Some instructors put up a very poor show at the blackboard, with small, spidery drawings and writing, which indicate a certain lack of confidence in their ability. Here are a few tips to bear in mind when using the blackboard. Always start with a clean board. If you are right-handed, have it on your left as you stand facing the class. Stand squarely before it and write with your right elbow held up well. Move the whole arm, instead of just the wrist. In diagrams, make your strokes from left to right, and from top to bottom, moving across the blackboard instead of standing at one side. Do not speak to the blackboard; the class will not be able to hear you and no notice will be taken of any thing said. There can be exceptions to this rule, provided what is said is in the nature of casual unimportant patter that is soothing rather than distracting. When you have finished the blackboard work, stand clear, and use a pointer.

Dark-coloured chalks show up very badly on a blackboard. Yellow is the colour most easily seen by the eye, then white, pink, red, green, blue, brown, purple. Use white or yellow for all writing on the blackboard, and a damp cloth for cleaning.

It is often useful to divide your blackboard into two by a chalk line down the middle, using one half to develop your blackboard summary of essential points, and the other half for general use, odd diagrams, calculations new words, which can be rubbed off as soon as you have dealt with the points. Some instructors use two separate blackboards.

A good tip for the instructor inexperienced in the use of a blackboard, or even an experienced instructor who has to build up a diagram requiring good spacing and arrangement, such as a radio circuit, is to draw a diagram carefully beforehand, then run a pencil line over the chalk lines, and rub the chalk off with a duster. The pencil lines remain, invisible to the class but not to the instructor. Your drawing will be carried out well without effort, and without having to consult the book or notes every few seconds when building up the diagram.

As a general rule, chalkboard work should be kept as simple as possible. Avoid spending a lot of valuable time (ten or twenty minutes) presenting a complicated diagram, even though well planned in advance. It is better to present this on a previously prepared chart, or on a duplicated handout. A valuable feature of the blackboard is that it can be used in conjunction with any number of other kinds of visual or audio-visual aids.

A small, light, three-ply blackboard, say 30 cms square is often useful in the course of practical work in the shop, or a piece of outdoor instruction for explaining a technical point to a group. It is better than the hastily produced back of an envelope or a cramped sketch done in pencil on the nearest piece of white paint! Keep a portable blackboard available for such occasions.

Blanket board, flannel graph or felt board

This is a useful teaching aid and can be either a dark blanket laid over a blackboard, or a piece of three-ply hardboard size A. covered with a dark-coloured rough-textured cloth. Whatever you want to show is written or painted, on cartridge paper, using felt pens of different colours, and then cut into the required shapes. These shapes are then backed with medium sandpaper, and they will stay securely on the blanket board in any required position if gently pressed to it. It is found that a good paper made rough on one side by roughing it up with fine glass paper will serve very well. You can build up diagrams, formulae, organisation charts, and develop summaries of your lesson, quickly and easily, in a striking and pleasing manner. (If you have to teach tactical play in games, the blanket board is invaluable.)

Plastigraph

An alternative to the felt board or magnetic board is the plastigraph which consists of a plastic surface to which Cellophane shapes will adhere securely. The shape or sheets can be conveniently cut, or written on with felt pens and are available in a variety of bright colours.

Magnetic board

Similar in effect to the blanket board, this consists of a thin sheet of iron or tinplate on to which shapes backed with small strip magnets will adhere. The strip magnets are sold in stationers' shops as they are now used for display purposes.

Also available for the use of the imaginative instructor are sheets of coloured magnetic material which can be cut into required shapes with scissors. Magnetic string is also available. Presentations of educational programmes on television make good use of this visual aid. Instructors could well emulate some of these professional presentations.

A very valuable feature of the feltboard, plastigraph and magnetic board is that they all require careful preparation and a planned sequence for their use and build-up.

The chalkboard alas can be and often is used as an unplanned scribbling block, as it does not demand prepared 'visuals'. One rarely sees the chalkboard in use during a television presentation.

Projected aids

Diascope or slide projector

This is a great asset and convenience in technical teaching because it can project slides or photographs of diagrams on the screen for a large number to see. Slides can be made from illustrations and diagrams in textbooks, using a small 35 mm single-lens reflex camera, and facilities for developing and printing. Transparencies can also be mounted in standard 2 in × 2 in slide holders. It is worth noting that slides of diagrams in black and white—or in colour using a colour transparency can be very easily made. It is not necessary to print, as the developed film negative can be mounted and projected.

A black and white diagram of course will appear as a white lined diagram on a black background, but this is no particular disadvantage.

A series of black and white slides, for example, can be made and projected in sequence to illustrate the various steps necessary in using a slide rule, for various computations.

Film-strip

This is a convenient way of showing a series of pictures, diagrams and captions in a planned sequence. A film-strip projector is in effect a small convenient diascope using slides joined together on film; it can be used also to show 2 in × 2 in slides. The advantage of slides over the film-strip is that the instructor is freer to vary the sequence and to make up his own lesson from a variety of slides.

A film-strip can easily be cut up into frames and each frame mounted in a standard 2 in × 2 in mount obtainable from any photographer. Many projectors are fitted with 2 in × 2 in slide magazines which can be focused and operated by remote control.

Lecturer's notes are often provided with a film-strip, but the temptation to read these notes verbatim should be resisted or the performance will be wooden and impersonal. Prepare your own lesson and use the frames on the film-strip in the order you think best to suit your purpose. Further, use all the teaching techniques as for an ordinary lesson, for example question and answer, tests to confirm learning, etc. The film or film-strip is an aid to a lesson and not the whole lesson.

Project in a room not entirely dark, otherwise contact is lost between the instructor and his class. When planning to project slides or a film to a group it is important to arrange for a suitable image size. This shoud not be too large, because, if you double the linear size of the projected image, you divide the luminous intensity of the image by four. Back-screen projection using a 'daylight' translucent screen and a mirror is recommended for showing film-strips and also has several advantages in projecting a film. Using this arrangement the room need not be dark, the class can take notes and the lecturer maintains contact with his class and has control over the projecting apparatus. The size of the screen, and consequently the image is normally about 75 cm square which provides a bright picture even in an undarkened classroom.

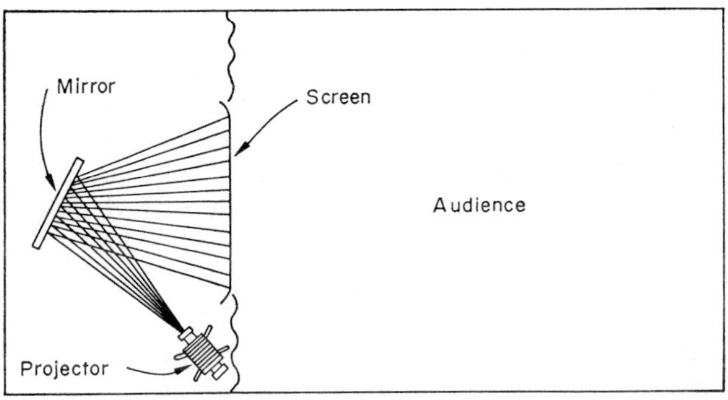

FIGURE 45 Example of back-screen projection

The geometry of projection

It is common in dealing with projected visual aids, for example a cine projector, to give a comprehensive table showing the picture size and performances of a variety of projection lenses from 25 mm to 150 mm focal length at various distances from the screen ranging from about 5 m to 30 m. The instructor can dispense with such tables and calculate the picture size from a simple application of similar triangles (see figure 45).

If l is the length of the longer side of a transparency to be projected (about 35 mm in the case of a standard slide), F is the focal length of the projector lens, and D the distance of the screen from the projector lens, then L, for practical purposes the size of the image (longer side), is given by $L = D\ l/F$, provided D is much greater than F.

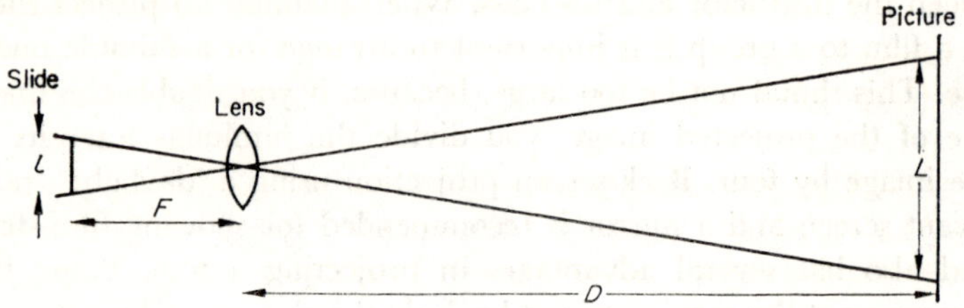

FIGURE 46

For convenience the whole projecting equipment can be behind a partition where it causes no distraction to the class, as in figure 45.

Episcope

This is an optical device for projecting an image of a strongly illuminated picture or a small surface on to a screen. The optical system of an episcope is often partly combined with that of a slide projector and the combination is called an epidiascope. The size of the picture that can be projected is generally limited to about six inches square. It is a great convenience for a lecturer to project his own material drawn on cards in coloured inks, or to prepare and project summaries or notes on cards in typescript.

The disadvantage of the episcope type of projection is that it can be used only in complete darkness, as the projected image is never very bright and the apparatus is usually cumbersome and heavy.

Instructors will find the episcope a great help in preparing charts or large-scale diagrams, and graphs, on sheets of paper up to the sizes A0, A1, or A2 (international paper sizes).

To make a chart from a diagram or graph in a book, the image of the material to be copied is projected on to a sheet of paper by means of the episcope. The paper, usually a good cartridge paper, is securely pinned in position on a chalkboard. The instructor then goes over the projected diagram or the parts that he requires, including the descriptive printing, using a pencil or a felt pen. This can satisfactorily be done freehand, and later touched up. Colours can be used as appropriate.

This procedure saves hours of work, and gives excellent results.

Overhead projector

This is a comparatively new training aid, which is of great value and convenience in supplementing chalkboard work or even replacing it (see figure 49).

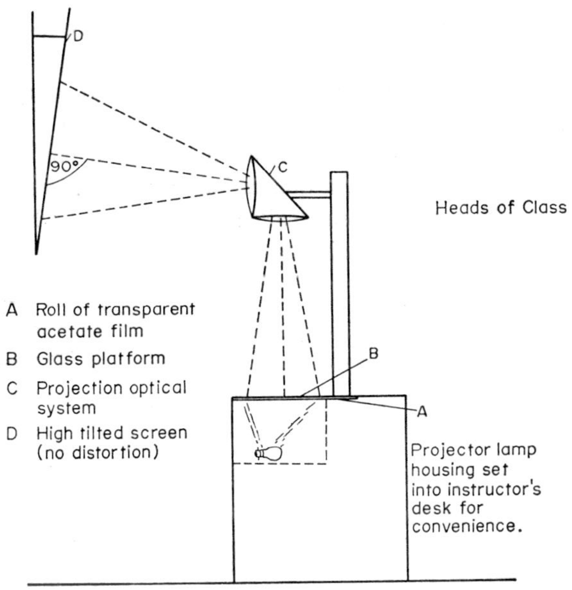

FIGURE 47 Overhead projector

Figure 47 shows how it is used, and why it is so called. It is in principle a transparency projector or a form of diascope. The transparency can be in the form of a roll of transparent acetate film ten inches wide, which passes over a glass platform about one foot square; the area of film thus spread on the platform is strongly illuminated from below, and forms the slide. Alternatively, square sheets of transparent acetate can conveniently be used instead of a roll. Using a 'chinagraph' pencil the instructor can then draw sketches or make summaries as if the film were a small blackboard. Transparencies of diagrams, formulae, summaries, etc., can all be conveniently prepared in advance, and a great deal of time can be saved in presenting material. A mirror and lens system projects in daylight an image about 150 cm square on to a screen behind and just to the side of the instructor who can face his class all the time. The instructor can draw and rub out as he talks facing the class and can use his pencil to draw attention to any point of his work. Colour can be introduced using a felt pen. To clear the screen you can turn the roller and bring a fresh section of film on to the platform. A previous drawing or set of notes can be recalled to the screen by just winding back. The roll can be used to store a course of lessons so that the teacher can allow any absentees to get an idea of what took place in the lessons missed.

There are two main forms of these projectors. In one, the light from the projection lamp is reflected out of the light box into the focusing head and through a projection lens by means of a large curved mirror; in the other, a quartz–iodine light source is used in conjunction with a special lens (called a Fresnel lens) which acts as a condenser. Both types of projectors are fan-cooled.

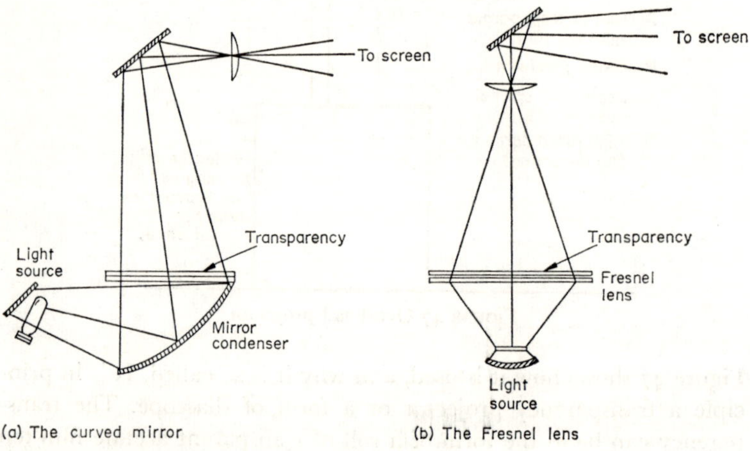

FIGURE 48

Some techniques

(a) *Pointing*—a pointer can be a pencil touching the transparency under discussion. The pointing object (pencil, pen, finger, etc.) should be rested on the projector stage to eliminate movement.

(b) *Chalkboard technique*—a grease pencil or felt pen saturated

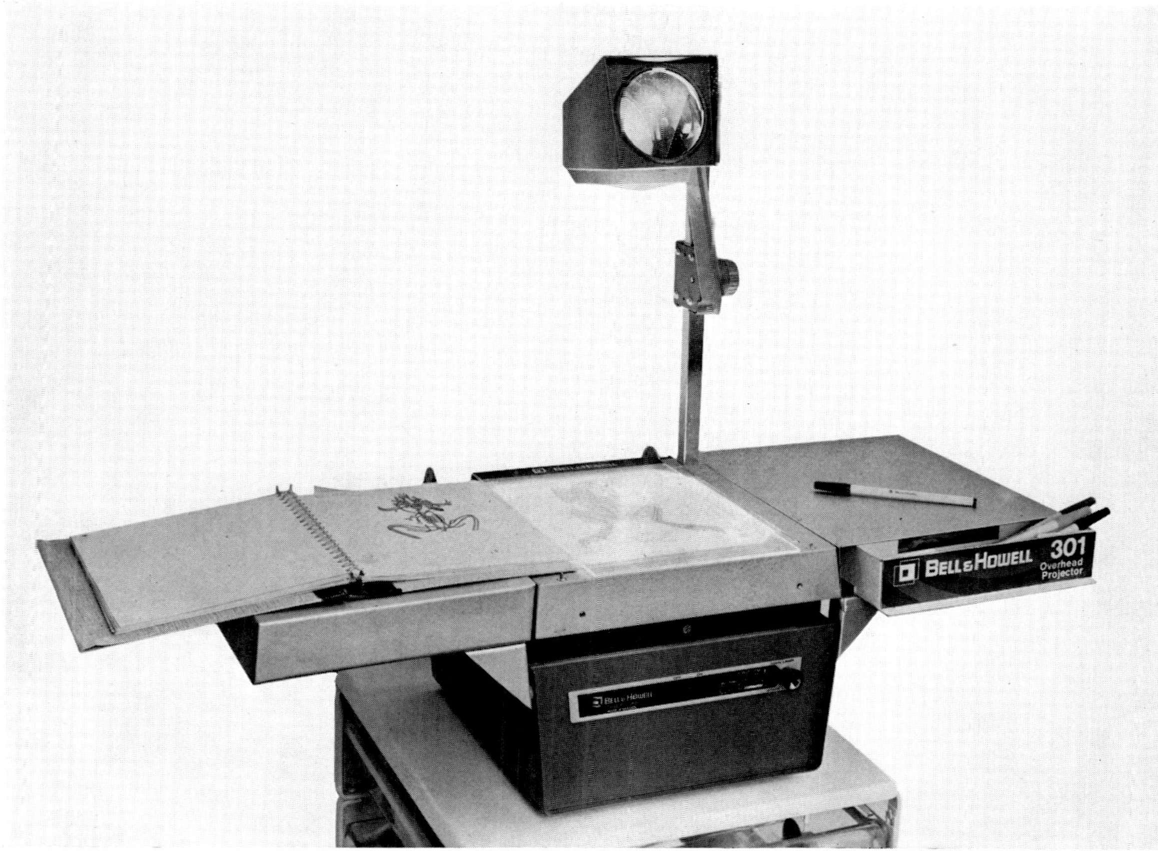

FIGURE 49 Overhead projector, showing binder for transparencies and overlays

FIGURE 50 A modern 8mm film projector, with automatic threading. Scene repeat. Combined optical sound and magnetic record and playback facility

with a water soluble ink is used instead of chalk and a clean, unexposed sheet of transparency film used in place of the chalkboard. Mistakes can be eliminated by wiping the transparency with a soft rag or even a moist finger. Tests or diagrams may be prepared in advance. Labels or other data can be added or removed.

(c) *Revelation technique*—an opaque sheet of paper is the simplest 'revelation mask'. Placed over all or part of transparency, it blocks areas you do not yet wish to reveal to the class. If you wish to cover material already revealed, while keeping material not yet discussed under cover, several strips of opaque paper or specially shaped masks may be used.

(d) *Silhouette technique*—any opaque object placed on the stage of the transparency overhead projector casts its shadow outline on the screen. Some shadow pictures could be triangles, circles, squares, rectangles, map outlines or lettering outlines. Biology teachers can show leaf specimens for comparison or identification by placing leaves side by side on the projector stage and casting their images on the screen. Physics teachers can show movements of iron filings in a magnetic field. The projector can also be used to project hydraulic experiments in a vertical plane, by lifting the projector through 90° and adjusting the focusing head.

(e) *Projecting transparent objects*—slide rules, triangles, compasses, and protractors in transparent form, can be projected onto the screen. Could be useful for instructing class on use of slide rule, etc.

(f) *Coloured transparent tapes and sheets*—transparencies can be supplemented with coloured transparent pressure-sensitive tapes. Graph work is especially suited to the use of this material. Coloured plastic sheets can be used to provide a full-colour background to a transparency.

(g) *Overlays*—information which lends itself to this type of presentation is information that can be broken down into component parts. A transparency is made of each component, and then the component transparencies are taped together on to a frame in proper registration.

(h) *Combinations of the above in more complex demonstrations*—Applications of the revelation of silhouette techniques combined with the projection of coloured transparent materials are the basis for a number of fittings that enable relatively complicated situations to be projected. Miniature science 'laboratories' have been developed and can demonstrate, for example, precipitation and effervescence of chemical reactions, movements of small animals in dishes of water, electrolysis of water, or the working of simple machines. Remember though, that opaque materials such as chemical precipitates will appear black on the screen.

Some advantages

i. The main advantage of the overhead projector is that the teacher is able to provide illustrations of greater breadth and depth than he can on a conventional chalkboard. By using a roll of acetate film (usually 25 cm wide by 15 m long) which is fitted on rollers at the sides of the projector, the teacher can draw a series of about 60 diagrams without the need to erase any of them. Each completed diagram can be moved across the screen to expose a fresh area of film on which the next illustration can be drawn. A review of all illustrations on the roll can easily be accomplished by rewinding the roll to its starting position. By this means either a teacher can illustrate his lecture with fresh diagrams, drawn *ad hoc* on each occasion, or, if he prefers, he can use a roll of illustrations prepared in advance and retained permanently.

ii. Since the illustrations are projected backwards on to a screen placed behind the teacher, he is able to remain facing his class and does not run the risk of losing contact intermittently as he does when writing on a chalkboard.

iii. The projector does not require a blackout; it can be used in the normal lighting of a classroom or lecture theatre.

iv. Although it is customary to use a screen the image can be projected satisfactorily on to a smooth, light-coloured wall.

v. The overhead projector is not limited to static diagram. It can show moving diagrams and models and can project many experiments as they are performed.

vi. It can make working with a small group more informal and writing for large groups less strenuous.

vii. Complex work may be prepared and checked in advance, thus saving valuable lecture time at the chalkboard.

viii. Transparencies can be made in a few seconds by a simple contact process—such as that done by the 'Thermofax copier' or a Verifax (Kodak) copier from a printed page. This means that many diagrams, tables, and formulae can be projected with far greater effectiveness, in daylight, than can be done with the episcope.

Some hints

1. When talking about a projected image look either at the transparency or at the class not at the screen. Point to the transparency, although some instructors prefer to point to the screen.

2. When you do not wish to use the projector even for a few moments, *turn it off*. The glare can annoy your audience.

3. Turn the projector off when you change from one transparency to another or while you are setting up objects on the stage of the projector. Disorganised movements on the screen are irritating and distracting.

4. When you use a transparency as a substitute for a chalkboard by sketching or writing on an acetate surface, consider using acetate sheets instead of rolls to make it easier to re-use the material. When acetate sheets are mounted by taping to inexpensive mounting frames they may be controlled more easily on the projector. Mount transparencies in a cardboard frame and store in a filing cabinet.

5. Writing with specially produced felt pens is preferable to using wax pencils. Do not, however, use ordinary felt pens as the inks are not transparent and so colours will appear black on the screen. Lecturers should use only *specially produced water soluble inks*—other media are virtually impossible to clean and costs of replacing acetate rolls become prohibitive.

Note-taking

Lectures or lessons cannot be thoroughly assimilated during the period of instruction, and a memory aid in the form of notes that can later be studied is therefore essential. Notes are also valuable for making references for further reading and study.

Controlled note-taking during specific pauses may be useful. An inquiry was conducted on the value of uncontrolled note-taking as distinct from a précis given to the class. Two equally graded classes of fifty trainees were the subjects. Class A were given précis for ten lectures; Class B took uncontrolled notes of the same lectures. The results of the inquiry is as follows.

	A		B
Précis studied and memorized accurately.	8%	Notes re-written, studied and learnt.	4%
Précis read but not memorized.	82%	Notes read through and polished up.	8%
Précis not read, but filed and available when needed.	8%	Notes not looked at after period, but legible and of a certain limited value for reference.	64%
Précis lost.	2%	Notes lost or illegible.	24%

It is tempting to draw various conclusions from this, but the most obvious one is that uncontrolled note-taking is not worth the time and loss of attention involved.

It is therefore generally unwise to allow apprentices, and classes below first-year university level, to take uncontrolled notes.

Laborious and time-consuming dictated note-taking can be a waste of time when the instructor has so many facilities for duplicating notes. These should be summaries of what has been studied and understood, and already presented on the chalkboard, feltboard or by the overhead projector.

For many trainees, however, the act of their writing something down is a degree of participation and helps learning. It is better than collecting a duplicated précis and never looking at it again!

Instructors taking an easy way out of their job may be tempted to dictate reams of notes in the belief that they are communicating their skills.

It is a deadly process.

A dictated lesson can be a process whereby information passes from the notebook of the instructor to the notebooks of the trainees, without passing through the heads of either!

A good plan is to write on the chalkboard the main headings and sub-headings as you go along, together with any new words of technical terms that need explaining. Many trainees will lack discrimination over what to write down, as they will be unable to discern what is important and what is not until the subject can be seen as a whole —at the end of the period.

Invite the class to fill in the framework from what they have remembered. Meanwhile the class can ask questions on obscure points to refresh their memories.

In preparing notes, particularly for duplication, see below, the following simple rules may help:

1. omit all that is not essential;

2. present the material in a systematic manner;

3. ensure that the level of the condensed material is appropriate to trainee's capacity to assimilate it;

4. provide the material at the right time.

It is better to withold your précis or notes till the end of the lesson, otherwise you encourage trainees not to attend to your own presentation with the summary before them.

Economy of space is important, so use graphs, diagrams, charts and tables whenever possible.

At regular intervals the instructor should collect and examine notes to see they are satisfactory and should give marks on a simple five-point scale, taking into account the correctness, neatness and arrangement of the material as well as the general presentation of notes including the standard of English.

An ability to write clear simple English is essential for aspiring technicians or technologists, and therefore training should include opportunities for trainees to practice both oral and written English. Good communication between people at all levels, craftsmen, technicians, technologists and scientists is vitally important in industry and means using language effectively and efficiently.

In technical colleges, special attention is paid to this aspect of a student's training by the appointment of English specialists. The technical teacher obviously cannot turn himself into an English language specialist, but he can insist on a high standard, and point out the dangers of ambiguity or lack of clarity in written or spoken work, in the hope that students will appreciate the importance of being able to say what they mean, and that they will strive to improve their powers of self-expression.

It is time well spent at the beginning of a training session to give students advice on how to plan their studies, where to go for information or help, and how to take notes and use them, for both formal classwork and practical work. A loose-leaf system is generally recommended as a loose-leaf notebook can grow or be modified as required.

Notes should be dictated verbatim *only* to confirm important features such as summaries or definitions. The effect of verbatim note-taking decreases with its continued use. Three points dictated may be remembered, thirty will be a waste of time, and the habit violates all the rules of good instruction. The issue of pamphlets and duplicated notes are of little value unless the student is very keen, and will *use* these handouts intelligently.

In favour of students taking notes on their own, one should remember that each has his own style and method and way of forming associations with words and ideas. His own notes are more intimate and personal than textbooks or duplicated notes.

As a rough guide, ten minutes in every forty could be used for note-taking during a lesson imparting knowledge in a technical subject.

Notes prepared by the instructor in the form of a précis of the material to be taught have two main uses—they guide the instructor and serve as a memory aid for the trainee.

For classes of average intelligence the tabulated précis should not contain more than 150 to 200 words. A longer précis cannot be memorised, and the trainee must be induced to read a précis to the end. It should be well aranged and clear even at a glance.

The issue of a skeleton précis which trainees may complete during the instructional period is a useful compromise between note-taking proper and the issue of condensed notes.

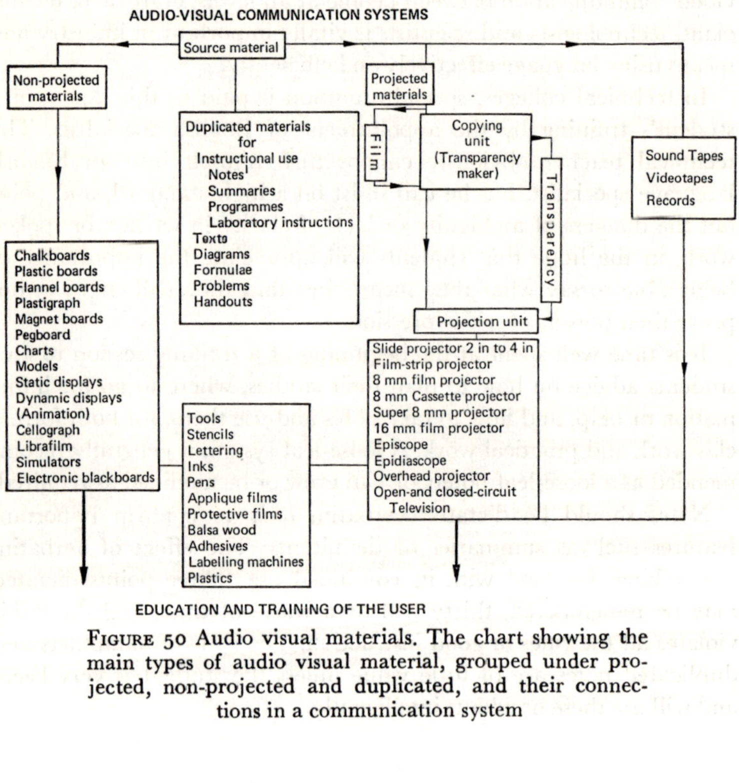

FIGURE 50 Audio visual materials. The chart showing the main types of audio visual material, grouped under projected, non-projected and duplicated, and their connections in a communication system

The skeleton précis is particularly useful in teaching technical subjects when diagrams can be printed, and the student must write his own explanations, names of parts and their functions.

Duplicated materials

One of the cheapest and most effective learning aid is the well-prepared duplicated sheet of paper. Handouts of various kinds can save a great deal of time for the instructor and the learner. Some examples are: lecture notes; summaries—these are specially useful when they correspond to the chalkboard or feltboard summaries; laboratory worksheets; diagrams; formulae; problems; tests.

One of the most convenient methods of duplicating material for classroom or workshop use is the rotary spirit machine which is a hectographic process duplication and can reproduce up to one hundred or more copies from typescript or the instructor's own writing and drawings. Various colours can be used, which is a great advantage in many diagrams.

Ordinary office duplicating machines are commonly used but these are mainly for typescript. Line diagrams can be reproduced, but the stencil has to be drawn with great care.

A wide range of methods and machines has now been developed to enable instructors to prepare duplicated copies of diagrams from books for class use. The subject is highly technical and has been given the name of reprography. The instructor should appreciate how such duplicated material facilitates his own work, and helps trainees.

Teaching machines

These are devices for presenting a programme of instruction for programmed learning. A brief account of this was given in chapter 8. There are several types of machine, including the following. Each of these can be operated in a variety of ways, using various electrical and mechanical control systems.

1. *Programmed textbooks.* These are in effect specially written and arranged books, through which a student can work, and in which the material is broken down into suitable steps each of which is tested and confirmed before proceeding to the next.

2. *Written response machines.* These involve the writing of short

answers on paper sheets, or on rolls of paper, or discs. Having written his response and passed on to the correct answers, the student cannot usually turn back to change his answers.

3. *Key response machines.* Responses here are recorded by pressing one of several buttons. They are generally for branching programmes and are much more expensive than machines for linear programmes.

A simple example of a teaching programme of interest to intructors can be derived from the multiple-choice test, with answers, in Appendix I.

A very simple and effective 'class response' machine can be made for a few pence, and used in a variety of ways.

Each member of the class is provided with a cube about 50 mm side with each of its six sides clearly marked A, B, C, D, ?, and blank respectively. A series of multiple-choice questions with four choices, A, B, C or D is presented to the class by means of an overhead projector, or on a duplicated sheet, or written on the chalkboard. Each member of the class then presents his response to the instructor, at the same time shielding the other five responses with his hands, so that only the instructor knows what each trainee's response is.

A trainee presenting the face marked '?' indicates that he has not understood the question, while the blank face could be used to show that all is well, after some explanation.

This simple cube (sometimes referred to as the 'Cosgrove' cube from its first reported use at the R.A.F. station there) effectively and pleasantly carries out the same functions as an elaborate console consisting of a central panel with rows of lamps electrically connected to six push buttons mounted on each of the trainees' desks.

Manufacturers of equipment for innovations in educational technology will naturally be glad to make these sophisticated but costly devices, which often do nothing more than the wooden cube.

4. *Audio-visual machines.* These involve the supplementary use of film, film-strip, microfilm and tape recorder.

5. *Computer-controlled machines.* The programme is fed into the computer and stored, so that it can be projected in the form of a cathode-ray image of a film, film-strip, microfilm or a combination of all three.

Systematic diagnostic tests such as fault finding and medical diagnosis can be computerised. That is a machine can give the answer to any question asked concerning the problem, and can detect and feed back any errors in logic that may have been made by the questioner in following up responses to previous questions. It can also point out irrelevancies or redundancies in any questions asked.

This has been used in the teaching and training of medical students, but the equipment is costly. An eminent surgeon, perhaps a little apprehensive of this impersonal approach, and the apparent absence of any sympathetic concern for the patient that a machine displays, remarked that, after all, the human brain remained the most effective and efficient all purpose computer, with the great advantage that it is mass produced by unskilled labour!

6. *Answer boards.* The student discloses correct answers by moving his answer sheet manually. It is not 'cheat proof' and has been described as an answer board rather than a teaching machine.

Catalogues of programmes for use with various machines are published in journals devoted to teaching aids, for example *The Visual Education Year Book*, published by the National Committee for Audio-Visual Aids in Education, 33 Queen Anne Street, London W.1.

There is some danger that teaching machines and other devices may fall into temporary disrepute at the hands of commercial gadgeteers, who will be tempted to take advantage of the enormous growth in the demand for educational technology, and put on the market equipment with programmes of dubious teaching value not based on a sound knowledge of the learning process.

Tape recorder

This can be a valuable teaching aid as we have already seen in chapter 8, when considering recorded sounds as a method of instruction (p. 173).

Museums

Visits to scientific museums can be valuable and instructive if well prepared and organised as part of a project. A technical college or training centre can form its own special collection of exhibits showing interesting pieces of equipment, models, samples of students' work, scientific specimens or even a small 'chamber of horrors' showing the consequences of bad work or neglect such as damaged tools, run bearings, pitted valve seatings etc.

The use of a reference library

Seldom in the training programme is time devoted to explaining the value and importance of a library to the trainee as an aid to his studies.

A well-equipped works library makes reference material readily accessible and supplies a real need. It would contain such material as textbooks; standard reference books; periodicals; digests; manuals prepared by the firm; employees' handbooks (on the firm and its products).

Many trainees, particularly apprentices in technology, may aspire to study at university level, where their success will depend to a large extent on their ability to pursue private study. Training officers should therefore ensure that trainees are instructed early in their courses on how to use a library as a source of information and as an aid to study. The co-operation of the librarian in this would be most valuable. He could help by giving practical demonstrations to trainees on the classification and recording of material. The library, just as any other aid to training, must be used skillfully, under expert guidance.

In the chart on p. 25 'How we learn through the senses', all the items listed are in fact training aids in so far as they can be used to improve communication and to help the learning process. In addition to the better known audio-visual aids, the following, although abstract, make a valuable contribution to the learning process, and are real aids for the instructor:

surprise	puzzles that challenge
humour	competitions
realism	anticipation

These have all been mentioned as part of the instructor's stock in trade in previous chapters.

As a tailpiece to this section and to the book, it is worth while to point out that you, as an instructor, are yourself a special kind of audio-visual aid, and can be by far the most effective!

Instructor's preparation workshop

It is a great advantage to have a special room where teachers or instructors can keep all they require to make charts, diagrams, models, etc. Such a workshop should contain carpentry tools; a good selection of drawing paper; poster paints; brushes; felt pens, coloured inks; varnish; wooden laths; nails, screws, nuts and bolts in assorted sizes; drawing instruments; a duplicator; a work bench and a drawing board. A 35 mm camera is useful for making standard 50×50 mm transparency slides or for making film-strips.

The UNESCO *Source Book for Science Teaching* contains many valuable hints for such a workshop, and for making the simple demonstration models and equipment specially appreciated in the developing countries.

Learning resources

In response to the demand for people qualified to be responsible for the day-to-day maintenance, preparation and use of audio-visual equipment of all kinds the City and Guilds of London Institute has planned a course of 48 hours' duration which results, for successful candidates, in the award of a C.G.L.I. Audio-visual Technician's Certificate.

All training officers and instructors should undergo a course on audio-visual aids so that the right aid can be selected for the right piece of instruction at the right time.

To recapitulate a few points

Figure 52 (p. 213) lists a number of objectives, methods of instruction, modes of learning, and means of testing, which may serve as a guide when preparing a course. The lists also provide a way of identifying a piece of instruction.

For example the instructor might plan a sequence as follows.

The object is to teach a skill (1.2) to apprentices (2.4) by means of a demonstration (3.3) with actual equipment (4.2) and handout notes (4.9) followed by a test on the sequence of steps to be followed (5.11) and a practical test (5.4) the shorthand description would be 1.2, 3.3, 4.2, 4.9, 5.4 and 5.11.

The lists are drawn up to show the great variety of elements that are available to an enterprising instructor who often feels that he must stick to a well-worn path 3.2, 4.1, and 5.1, characteristic of the schoolroom of a past era.

A brief calculation will show that, having decided on the broad object of the instruction and the trainee target area, then there are about 17 methods, 22 kinds of learning aids and 15 ways of assessing. More than one method can be incorporated into one piece of instruction and several kinds of learning aids can be used. The means of testing can also be more than one kind.

The possible number of combinations taking only one element for each group is $17 \times 22 \times 15 = 5\,610$. Not all these will of course be suitable combinations, but the lists may provide material for thought, and may tempt the more conservative instructors to break out and try out some sound combinations using different methods, employing some new learning aids, and means of assessment.

WHAT IS INVOLVED IN TEACHING AND LEARNING

1. *Main objective of training*	2. *Trainee target*	3. *Method of instruction*	4. *Mode of learning (learning aids)*	5. *Means of testing*
	1. vocational school	1. lecture	1. actual equipment	1. written paper (essay type)
	2. pre-apprentice	2. lesson	2. models	2. continuous assessment e.g. question and answer
1. knowledge	3. operatives	3. demonstration	3. simulators	
	4. apprentices	4. practical work	4. handling things	
	5. works based	5. tutorial	5. discovery learning	
	6. college based	6. module	6. museums	3. multiple choice
2. skills	7. craftsmen retaining	7. discussion group	7. chalkboard	4. practical test
	8. upgrading craftsmen	8. syndicate discussion	8. charts	5. open ended experiment
	9. adult workers	9. projects	9. feltboard	6. oral test
	10. technicians	10. lecturettes by trainees	10. magnetic board	7. complete the sentence
3. attitudes	11. special skills	11. playlets–role acting	11. slides	8. supply the missing word
	12. university graduates	12. games	12. overhead projector	9. name the parts
	13. size of audience, class, or group	13. simulation	13. animated diagrams	10. complete the diagram
		14. brains trust	14. film loops	11. state the steps
		15. film—20 mins or more	15. closed circuit television	12. matching lists
				13. speed tests
		16. instructional visit	16. textbooks	14. interviews
		17. teach team	17. library resources	15. test by discussion
		18. competitions	18. hand-outs, notes	
			19. work sheets	
			20. tape recordings	
			21. programmed learning	
			22. teaching machines	
			23. computor aided learning	

11

Summary of the Main Points Concerning the Instructor's Job

This is in the form of brief notes for beginners who find themselves 'thrown in at the deep end'. (It was originally prepared to help inexperienced graduates, etc., who had volunteered to teach in developing countries.)

Attention through interest

You cannot teach a person unless he wants to learn. One of the teacher's jobs is to encourage a trainee to want to learn—in contemporary jargon, to motivate him.

Teaching is helping people to learn.

Importance of names

Have a seating plan made before you enter the class if possible and keep it before you so that you know the names of the students and where each one sits—preferably in the same place for each of your lessons. Try always to refer to a trainee by name.

Study what you have to teach

Study the material you have to teach. Get a clear idea of standard to be achieved and what the class knows already.

Seven stages in teaching a technical subject

1. Object
2. Preparation
3. Presentation } (Communication)
4. Reception }
5. Assimilation
6. Assessment
7. Feedback

1. *Object.* Your first job is to get the object of your course and each particular lesson quite clear. This may mean discussion with your training officer. The object of each lesson must be definite and limited. The overall object of training (better production, happier men on the job) must always be kept in mind. Allow interesting and stimulating diversions, but do not be led astray. Invite students who want to go off on a sideline to meet you after the lesson.

2. *Preparation.* This stage determines the success or otherwise of your lesson. The more you prepare, the more effective is your teaching. It includes the following.

2.1. *Selection of the main points to be taught.* The subject matter to be taught, its *selection* and arrangement into a proper sequence is generally laid down by the syllabus. A common mistake is to attempt to impart too much information at one time. It is better to teach a little well than to attempt a lot superficially. Basic principles are more important than details. Have your own priorities and go for the things the students *must* know, relegating to second place the things that it would be nice to know if there were plenty of time. Keep up to date. What was relatively unimportant five years ago could be very important today and vice versa.

2.2. *Sequence in your lesson.* Useful rules. Start within the student's experience, that is with familiar appliances and everyday experiences. Proceed from the known to the unknown, the simple to the complex, the concrete to the abstract, from the particular to the general. Start with the whole, tell your students what it is all about, then analyse it into parts and synthesise them back into the whole.

2.3. *Method.* Encourage the class to find out things for themselves and to discover a principle under guidance. This is better than the practice of stating a principle or rule and then setting about finding examples or 'proving it'. In teaching any kind of practical skill and emphasis must be on 'learning by doing', not on 'talk and chalk'.

2.4. *Timing your points.* Remember a class cannot normally concentrate on forty minutes' non-stop talk. The class is fully with you for about ten or fifteen minutes; after that attention is liable to wander, unless there is something to do or to look at, or a two-way traffic of questions and answers. Attention generally returns towards the end of the lesson (probably in anticipation of release from the class); use this period to punch home your main points, or to go over your blackboard summary.

2.5. *Find out what the class knows already.* It is boring to go over ground already adequately covered, and disastrous to assume a previous knowledge that is not there. The students' own version of what they know is often unreliable. The best way is to find out just what the class knows at the beginning, by questions or a short test consisting of a large number of short questions, or a simple 'alternative-answer' type of test. This will help you plan your first few lessons and give you some assessment of the knowledge in your subject of each of the students.

2.6. *How fast can I go?* In deciding this you have to be guided by certain psychological principles governing the learning process and assimilation. In general it is better to go at the pace of the bottom half-dozen rather than the top half-dozen. The disadvantages of going too slowly for the top students can be offset by giving them more advanced work on their own from time to time.

2.7. *Notes on audio-visual aids.* We learn far more through our eyes than through our ears. Do not talk too much. At the preparation stage of your lesson plan to use apparatus, real objects or models, charts, posters, film slides, blackboard, etc. Good visual aids focus attention, stimulate interest and make teaching and understanding easier. Choose the aid best suited to your purpose.

Models and three dimensional representations are generally to be preferred to pictures and diagrams. Visual aids should help you and your class, they should not dominate the instruction or disturb the class–teacher relationship. They should be simple with all non-essential details and lettering eliminated. Two or three primary colours are useful provided they separate parts and add clearness. Concentrate on one or two essential parts at a time and build them up into a whole. Use the real thing when ever possible.

Visual aids should be shown to the class at the right moment, and then covered up or put away. As a general rule do not use them as decorations, they become distractions and invite inattention. Some posters and wall charts, however, are specifically designed to rouse interest or to summarise a great deal of information and are intended to be on continuous display, but the display should be in workshops, in corridors, or at the foot of a stairway. Charts are specially useful when shown in sequence (clamped together) to illustrate the systematic 'build-up' or development of a topic. These can quite easily be made by the teacher using the modern methods of felt pad 'pens' available in several bright colours.

Visual aids' workshop. A great help is a small room or corner set apart where you can prepare teaching aids. The main requirements are a stock of large sheets of white paper 30 in × 30 in, poster paints, brushes, felt pad pens, coloured inks, wooden laths, basic carpentry tools, a duplicator, a 35 mm camera which can be used to make slides.

Blackboards or chalkboards. Should have a good surface. Blackboard paint, or even Indian ink, will often renovate an impossibly grey or worn surface. Show confidence by clear writing and diagrams. Do not speak to the blackboard, except for a piece of occasional unimportant patter. When you have finished at the blackboard, stand clear and use a pointer. Dark coloured chalks show up badly. Yellow and white are colours best seen by the eye.

When you have to build up a complicated diagram or table which requires care in spacing or arrangement, a good tip is to draw the diagram carefully beforehand on the blackboard *in pencil.* Pencil lines are invisible to the class but not to you. Your performance on the blackboard will be impressive and you will not have to consult your notes or textbook every few seconds.

The blanket board. This can take the form, for example, of a dark blanket laid over a blackboard. Whatever is wanted is written or painted on cartridge paper, and then cut into shapes which can be backed with medium sand paper, so that they will stay in position when placed on the blanket board.

2.8. *Further suggestions on preparation.* In teaching a skill such as soldering or riding a bicycle the most effective allocation of time has been found to be: explanation, 10 per cent; demonstration, 25 per cent; practice, 65 per cent.

You have to prepare not only the *matter* to be taught, but the *minds* of your students so that they are receptive and are ready for assimilating the new ideas. A few minutes' revision of previous lessons are often well spent. Finally, prepare all your materials, apparatus, demonstrations, diagrams, models, film-strips. Are they all in good working order?

3. *Presentation.* Having prepared your lesson the next phase is to 'put it over'. Transmission by the teacher and reception by the class is a unified operation or a process of good contact and communication between teacher and taught, with participation by both. Here are a few points on presenting your lesson.

You can infect the class with your own attitude and your outlook and enthusiasm. In this respect you are the most effective audio-visual aid in the classroom.

You must know what you are aiming at and give a sense of confidence to the class that you are achieving your common purpose.

You must have the feel of the class and be sensitive to its needs.

Know your subject and your class, but do not dazzle the class with your advanced knowledge. Use the word 'obvious' sparingly. Your lesson should run smoothly from beginning to end with no hitches or breakdowns.

The class should be comfortably seated so that they can all see you; the light should not be in their eyes.

Start and finish on time.

Vary the method of your approach until you are sure the class is all with you.

Competitions between sections of the class and puzzles challenge the class and stimulate interest.

Watch the faces of your class. They generally show whether you are going too fast or too slow.

3.1. *Your manner in presentation.* Note the following.

Voice. Speak clearly and distinctly in your natural voice but loudly enough to be heard at the back of the class. Vary speed, tone and emphasis. Get a tape recording of your own voice in action, root out your faults. Deliberate pauses produce expectancy but a series of 'ers' or 'ums' to fill in time advertises that you are stuck for words and irritates the class.

Eyes. Look at your class, let your eyes travel round it and let students feel you are speaking to them individually; do not speak to the clock at the back or the open window or the blackboard.

Mannerisms. Do not do anything which distracts attention. A mannerism that easily develops is to keep asking, 'Do you follow me?' or a similar rhetorical question which, by its frequency, becomes worthless. It is better to ask a series of specific questions to test if the class is following.

Stance. Stand easily on both feet—do not sit (unless conducting an informal discussion), lean or slouch. Your own appearance can be distracting if you dress untidily.

Teachers are critical of their students but often fail to realise that students apply a far stricter and severer standard in judging their instructors.

Attitude. Do not have favourites or strong dislikes. Humour is excellent for maintaining good relations but use it only when appropriate to the occasion.

A student appreciates any sort of personal attention you can give him. He is stimulated by the awareness that you know his good points as well as his bad ones. Best teaching results are obtained by handing out more praise than blame, but do it with discrimination and to individuals by name rather than to the whole class indefinitely.

A sound piece of advice in dealing with people is to put yourself in the position of the other person and see through his eyes.

If you are asked a question you cannot answer say, 'I'm sorry I don't know, but I will find out.' (Then find out!)

4. *Reception.* Effective reception by a student follows the process shown below, starting with the student's desire to learn. This may be weak at first, but the teacher aids the process at all points in the cycle shown below.

4.1. *Steps in reception.* These are shown below.

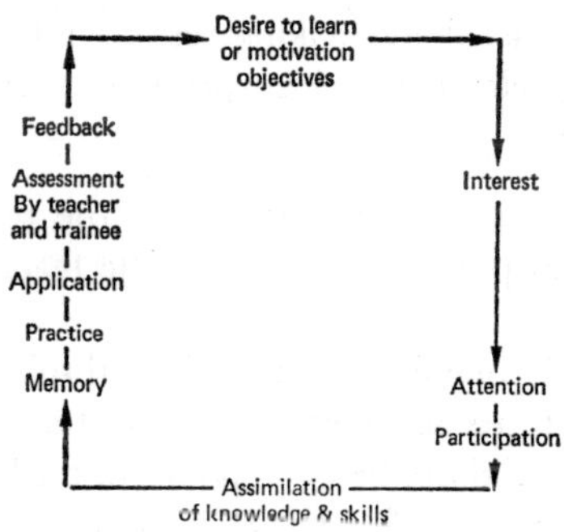

4.2. *Competition* is a strong incentive to effort and should be wisely used, for example by dividing the class into two parts and arranging competitions. This is more satisfactory than competitions between individuals which tend to make the brighter boys cocksure and discourage the weaker ones. Rivalry is ruined if the difference in standard between the competing groups is too great. Vary the competitions so that everybody has a chance to do well at something.

4.3. *Progress-reporting* is an essential part of the teacher's work. Students, in fact people of all ages, work better if they know the results of their efforts. (This principle is judiciously used in teaching-machines.) Remember praise and rewards are better than blame and punishments.

Keep the class as active as possible and let the class learn by doing, not merely by listening or looking. Students should be cheerful, co-operative and enthusiastic over their work. These qualities are largely caught from the teacher. They are not topics that can be taught.

5. *Assimilation.* This is an essential and all-important phase in the learning process and corresponds to digestion in eating. A hungry man eats, digests, absorbs food and the products are stored in the system ready for use. A motivated learner attends, assimilates, remembers and stores in his mind ready for use. As cramming is bad for the digestion, so it is equally bad for assimilation and produces disastrous results in either case.

5.1. *Meaning and understanding.* It is essential to teach what is meaningful and leads to understanding. We remember best what we attend to with full interest and what has meaning, is understood, and used successfully.

Assimilation is assisted by making lessons interesting, vivid, impressive, by using elements of surprise, curiosity, realism, class activity—competition.

5.2. *The use of questions.* Questions by the teacher and by the class are an essential part of any lesson. The teacher asks two types of questions.

1. To make each member of the class think. In this category questions are used to rouse interest or curiosity, or to lead the class by planned steps to build up the main points of the lesson.

2. *To test and confirm what has been taught.* Questions here are used to find out what the student already knows, to check whether the pace is adequate and prepare for the next advance. Your skill in asking and answering questions will largely determine your success as a teacher.

5.3. *General hints on questions. Name last.* Put your question first to the class, and then let your eye rove over the whole class. Pause, then name the student you wish to answer it. This makes it necessary for everyone to listen to the question and to think of the answer in anticipation of being asked. Suppress any tendency for students to call out the answer without being asked.

Spread. Invitations to answer questions should be spread evenly and fairly throughout the class.

Phrase carefully clearly, and in simple language.

Wrong answers. Can be valuable as they often indicate a weakness in your teaching. Use wrong answers to clear up hazy points, and to plan future revision.

Easy steps. There should be no big jumps in your reasoning, especially in teaching mathematics. Questions must be clearly put and asked in logical sequence. They should be such that a student can work out or discover the answer. Asking questions should never become a guessing game and if you ask silly questions you will get silly answers.

The answer must be heard. If necessary and sufficiently important repeat answers so as to make sure all the class has heard.

Avoid questions that have an obvious answer. They waste time. For example, 'What is a starter used for?'

Avoid trick questions. 'How many cylinders has a four-stroke engine?'

Avoid questions which the class cannot reasonably be expected to answer. For example, some formula that the class has not met before.

Avoid questions which test powers of expression, particularly when they relate to a practical skill, for example, 'How do you keep your balance on a bicycle?' You cannot test skill in anything by means of questions. The test for a skill must be a practical one.

Avoid questions with a 50–50 chance of a student getting it right, for example those that can be answered by 'yes' or 'no'.

Avoid pointless general questions which beginners often fire at the class such as: 'Do you follow me?' or 'Do you see what I mean?' Rarely will anyone have the courage to say 'no' and even if one student does say 'no' you do not know how general the situation is. It is your job to ensure that the class is following you by specific questions spread round the class. It is worth while preparing a dozen useful, well thought out questions for each of your lessons.

Encourage questions from the class at all times. Those which anticipate later work can be briefly answered but should be written down and taken up fully at the appropriate time by saying, 'Thank you for raising the point which we shall deal with next week.'

5.4. *Note-taking.* Students should take notes of the essential points of a lesson to help them in assimilating and recalling what has been taught. They should be written in a suitable form for future study, and contain references for further study and reading. Controlled note-taking in classes during specific pauses is often useful and is generally to be preferred to uncontrolled note-taking. You should inspect or mark class notes from time to time.

A good plan is to write on the blackboard the main headings and points as you go along—together with any new words or terms that need explaining. The class can fill in this framework for themselves at a set time. Indiscriminate note-taking distracts their attention from you. Notes should be dictated verbatim *only* to confirm important summaries or definitions.

The issue of duplicated notes or summaries has value particularly where the class is very keen, and will study them intelligently and with understanding.

A student's own notes—built round the framework or skeleton précis you have provided are of value. They are more intimate and personal than textbooks or duplicated notes.

You should prepare your own notes with a twofold purpose in mind: 1. your own guidance; 2. your summary and précis for note-taking.

5.5. *Libraries.* Encourage your students to use a library if there is one, and to look things up for themselves.

5.6. *Ending your lesson.* Use the last five or ten minutes to ensure maximum assimilation and to test how effective you have been in achieving the object of the lesson—questions are useful at this stage.

End with a brief summary or by going over the blackboard summary you have built up during the course of the lesson and which the class should have written down as notes.

6. *Assessment.* For the class a test provides something to work for and enables students to know how they are faring and what their weaknesses are. For the teacher, tests gauge the effectiveness of the teaching and his weaknesses, the progress being made by individuals,

and how much revision is necessary. They should, as far as possible, be designed to give confidence, encouragement and a sense of achievement.

A test should be a means of consolidating the position gained and of preparation for the start of new work and not an end in itself. There should be no slackening of effort after a test.

Give students the results of tests within a day or two. A long interval between test and results reduces its value.

Questions which call for an essay-type answer are easy to set but they test powers of expression rather than knowledge or ability to think and are difficult to mark.

Many teachers tend to regard marks with an undeserved reverence; in all but vigorously standardised tests, they should be regarded as merely a convenient means of arranging the examinees in order of merit. Examiners vary widely in marking the same examination papers, but they are much more likely to agree reasonably well over the order in which they place examinees.

A good examination is one designed specifically to test the abilities required to be tested which gives a good spread in the distribution of marks, so that the top third of the class is easily distinguishable from the middle third and the middle third from the bottom third.

6.1. *Some practical points in devising, conducting and marking tests* are as follows.

Large number of short-answer questions, rather than a few questions that require long answers.

Questions clearly worded. Have a clear idea of the answer you expect for full marks when you set the question, but give marks for good understanding of the subject and independent thinking.

In the alternative-answer type of test have the answer sheet gridded and a stencil cut with the correct answers marked on it, so that marking is automatic and independent of the personal judgment of the marker.

A test of skill must be practical. Test accuracy first, speed later.

Test knowledge and understanding, not power of expression.

Set some tests in which students can use their notes or text books to see if they understand the principles and can use them. This will assess understanding, not mere book work or factual knowledge.

Good seating arrangements, candidates well spaced, no serious distractions.

Have everything ready: paper, pencils, instruments, apparatus and instructions.

Be punctilious over the timing and indicate the passage of time at reasonable intervals. Allow no talking or communication between candidates.

Use a red pen or pencil for marking, as this helps the eye to pick up the marks more easily for totalling.

Give results to the class the next day if possible and use a 'post-mortem' to teach or confirm.

6.2. Teachers are often discouraged by an apparent but serious slowing down in the rate of progress either in the class or in the individual frequently due to what is termed 'saturation'. This is a kind of mental indigestion, one of the causes being that intake has been too rapid and the new material and ideas not assimilated. The effect is a 'plateau' in the learning curve. Saturation sets in after about ten minutes of fast and continuous lecturing particularly when a great many facts are being disseminated.

Lecturing with no breaks for audio-visual aids, demonstrations or questions has no place in the classroom. Ten minutes is as much as any class can attend to a continuous outpouring of facts uninterrupted by demonstrations, questions, class activity or visual aids, and forty minutes is a satisfactory limit for a well-prepared lesson.

Have an open mind and be ready to try out under ideal conditions any new and stimulating idea and teaching aid such as tape recordings, films, slides, television, teaching machines, visits to power stations, or other works or technical colleges.

7. *Feedback.* Good teaching promotes understanding and interest and a keener desire to learn. Good assessment both of your own efforts and those of the class reveals the effective and weak parts of teaching and learning and the results of assessment should be fed back into your next series of lesson plans.

Appendix I
Example of a Multiple-Choice Test, with Answers and comments

Instructions

Time allowed: 10 *minutes*

Read the following carefully.

1. All answers should be made on the separate answer sheet given to you, so do not write anything on this test paper.

2. The answer to each question is to be given by putting a ring round one of the letters A, B, C, D, opposite whichever statement you consider the best.

3. If you do not know the answer to a question, do not write anything against any of the alternatives. You will be penalised for wrong answers of guessing.

4. Work as quickly as you can.

5. As soon as the examiner says 'stop writing' put your pencil down on your desk.

Test

1. A good instructor when lecturing a large class should :

A walk about continuously;
B stand at the back of the class;
C stand in front of the class and move about occasionally;
D make himself comfortable by sitting down.

2. A good instructor should :

A speak as loudly as he can;
B speak in his natural voice with plenty of variations in speed and tone;

C speak very slowly and avoid variations in tone;
D try to imitate the style of the best public speaker he has heard.

3. You have an odd half-hour with your class. So you would:
A Tell the class to lose itself quietly so as not to disturb the rest of the establishment;
B set the class to read their notes;
C put a good apprentice in charge and relax in the staffroom;
D run a brains trust or similar game on the work of the past few weeks.

4. Demonstration models should:
A not be displayed at all in the classroom;
B be displayed prominently throughout the course;
C kept out of sight till wanted for actual use;
D shown only at the very end of the course.

5. An instructor when instructing should:
A have his lecture written out in full and use it;
B use a good textbook as his lecture notebook;
C use a few headings and sub-headings;
D use no notes at all.

6. An instructor should tell a funny story:
A whenever he can think of one;
B when appropriate to the occasion;
C not at all as it is out of place and interrupts instruction;
D whenever the class looks like going to sleep.

7. New technical words should:
A not be explained at all, as the meaning only comes with use;
B be used first and explained later;
C be written on the blackboard first, explained and then used;
D be explained and not used for a few days.

8. Notes of each lesson are most effective when:
A dictated to the class by the instructor;
B taken down during the lecture uncontrolled by the lecturer;
C copied out from the textbook as convenient;
D taken from the blackboard as headings and main points at the end of each natural division in the lesson.

9. We obtain most of our experience through our :

A ears;
B eyes;
C hands;
D nose.

10. When teaching a class the instructor should :

A confine himself to lecturing;
B appeal to as many senses as he can;
C lecture most of the time and appeal to the senses of sight and feeling sparingly;
D use the sense of sight at the end of the lesson.

11. In an initial piece of instruction on the lathe you would :

A give the class a good lecture using a diagram of a lathe;
B give the class what they ought to know about it first, and keep the actual lathe out of sight till the end;
C let the trainees see and handle the lathe under your supervision during the first period of the course;
D arrange for the class to pick up on their own what they can about turning for a few days and then come to the class for instruction.

12. Questions should be :

A asked mainly of the bright trainees;
B asked mainly of the dull trainees;
C asked mainly of the trainees who are too nervous to ask questions;
D spread evenly among the class.

13. Questions should be asked chiefly :

A to keep the class awake;
B to catch the lazy ones;
C to obtain the active participation of the class;
D to encourage the best ones.

14. A small class of men under instruction should be :

A encouraged to ask questions;
B told to ask questions individually after the lesson is over;
C told to write questions down and pass them up;
D told not to ask questions but to listen carefully.

15. When asking a question the instructor should :

A address the question to the whole class and allow the man who knows to answer up smartly on his own initiative;
B address the question to one man;
C address the question to the whole class, pause, and then select one man by name to answer;
D address the question to the whole class, pause, and give the answer himself to save time and avoid wrong answers.

16. If you are asked a question you do not know the answer to, you would :

A skilfully turn the topic on to a point you know the answer to;
B say, 'I will deal with that point to-morrow';
C say, 'I am sorry I do not know; but I will find out';
D say, 'You ought to know that. Look it up for yourself and tell me to-morrow.'

17. In a class of twenty apprentices an examination showed the following results :

one man obtains 95 per cent;
nine men obtain between 90 per cent and 95 per cent;
seven men obtain between 80 per cent and 90 per cent;
three men obtain between 70 per cent and 80 per cent.
What is a reasonable conclusion from these results :

A the class is a good one;
B The examination questions were too easy to test the class properly;
C the question paper had leaked out;
D the examinees have copied.

18. A class, 'A', of apprentices obtained an average mark of 58 per cent on passing out. Another class, 'B' on a similar course and exam., six months later obtained an average of 52 per cent. What is your conclusion?

A class A is better than class B;
B the examination was harder for B than for A;
C class B was not as well trained as A;
D no conclusion is possible.

19. You ask to see a report on an apprentice, and his marks in five successive class tests are 46 per cent, 49 per cent, 52 per cent, 53 per cent and 55 per cent. From this information what would you conclude?

A the apprentice has improved;
B he should have shown more improvement;
C from these marks no conclusion is possible;
D the general standard must be low.

20 The best time of day for a difficult piece of technical instruction is:

A in the evening after a football match;
B the first period of the morning;
C the end of the morning;
D just before tea.

21. You want to get the best out of a good class of men on a 'national certificate' course. You would:

A maintain and publish class and individual 'progress reports';
B tell the whole class from time to time that they are much poorer than the last lot;
C tell the whole class that they are much better than the last lot;
D 'take it out of the class' in every possible way, to keep them up to the mark.

22. You are responsible for making up classes. Which policy would you adopt?

A change men from one batch to another as frequently as possible to prevent the formation of 'cliques';
B change men as little as possible to stabilise groups;
C change a few best types frequently;
D change a few worst types frequently.

23. During a course for men selected for training prior to promotion to a higher grade, the object of a personal interview should be:

A to keep the man in suspense till the end of the course;
B let him know where he stands and what his chances are;
C to tell him in general terms that he must 'get down to it' or 'pull up his socks';
D to let him feel he is bound to make the grade no matter what he is actually like.

24. You have an apprentice who has gone right off in his work on a course. You would:

A call him for a personal interview and find out what is wrong;
B ignore him and let things right themselves;
C give him a severe reprimand and warning in the presence of the class;
D get every instructor and foreman in the firm who deals with him to see that he does not slack.

25. You have been asked to give a course to a newly formed class of men. The *first* thing to do is:

A study the subject hard;
B find out what the men know;
C find out what the object of the course is;
D talk to some men who have already done the course.

26. You have a class of fifteen men sent to you from another branch of the firm for a special technical course. You would:

A assume they know nothing and start from scratch;
B assume they know what you think they ought to know;
C hold a short test on the first day and find out;
D find out what they know as you go along.

27. You have an industrial film illustrating a part of your course. You would:

A keep it till the end of the course;
B keep it for an odd morning or Friday afternoon;
C put it on when the interest seems to be at a low ebb;
D use it in a place that suits the programme of the course.

28. You have forty-five apprentices to be divided into three classes of fifteen each. They are all to be on the same course. You would:

A divide them into classes alphabetically and tell the instructor to go at the pace of the best in the class—the others will follow on;
B divide them into classes to ensure a good mixture in each class and go at the pace of the slowest. The best man can take it easy if the pace is too slow;
C set a short test to grade the men and form three classes according to their ability to learn. Each class will then have men of

about equal ability and lose the stimulus (if any) of the best men;

D leave it to the office to decide, and not bother with such details as they are unimportant.

29. You have to plan a syllabus for a piece of technical instruction with practical skill as the main object. You would :

A spend most of the time demonstrating, as the practical skill will come when the men get on the job;

B allot about 20 per cent of the time to practical work, the rest of the time to explaining;

C allot 50 per cent of the time to practice;

D allot more than 50 per cent of the time to practice.

30. A teaching machine is a good and useful device because it :

A reduces the work of the instructor and could replace him;

B reduces practical work to a minimum;

C works to a graded programme in which the students participate;

D encourages the weaker students.

Answers and comments

1.A No. A distracting mannerism.

B No. Contact with the class is lost.

C *Yes*. Natural and best for maintaining contact with the class.

D No. Induces a sence of relaxation, and shows a lack of enthusiasm on the part of the instructor.

2.A No. It is very trying for the class to be shouted at, and a strain on the instructor.

B *Yes*. Correct provided the whole class can hear you.

C No. Monotonous and boring.

D No. Produces unnatural affectations.

3.A No. A very lazy way out and will not convince the class you are keen on their progress or your job.

B No. Also a lazy way, and will not ensure any useful work is done.

C No. Shows a lack of interest, and such attitudes are communicated to students.

D *Yes*. This means an effort, but it will be worth it.

4.A No. Demonstation models are valuable aids to teaching.

B No. Likely to become distractions, and cause attention to wander from the point you are dealing with.

C *Yes.* Ensures full attention at the right moment, decided by you, and removes the possibility of distraction.

D No. Models should be used at the right moment in the lesson.

5.A No. Allows no interesting variations, and will not deal with students difficulties as they arise.

B No. Students can and should use a textbook but soon discover that you are taking an easy path, and making little effort in your preparation.

C *Yes.* Ensures proper emphasis of essentials, good chalkboard work, admits variations in teaching methods to suit individuals or situations as they arise, and facilitates note-taking.

D No. Dangerous except for a very experienced instructor. Instructors should be well prepared. Notes of questions and summaries are important, both for you and the class.

6.A No. More the task of a comedian. Students will tend to pay more attention to the jokes than to the instructor.

B *Yes.* Helps to foster good class–teacher relationships, adds interest, and when skilfully done, aids memory.

C No. Need not be out of place, and if appropriate can aid assimilation.

D No. May be useful for awakening a class, but why are they going to sleep? Inattention is the teacher's sin. Good preparation, presentation and class participation with question and answer are better expedients.

7.A No. Time consuming and bewildering.

B No. Introduces an unnecessary period of mystification.

C *Yes.* Introduces a sense of sight as well as sound. The new word is given a good introduction, with spelling, prior to proper use. Use the word frequently after explanation and get the students to use it.

D No. Will not aid memory.

8.A No. As a general practice becomes an easy way out for a lazy instructor. A poor class will often accept this method but a good class will protest against it.

B No. Can lead to a perpetuation of misconceptions. Students often have a wrong appreciation of what is essential or important, and may miss important points by being engrossed with the tasks of making notes.

C No. Denotes lack of control, supervision or even interest on the part of the instructor.

D *Yes.* The essentials are thus assured. The student can later work on the notes and make them his own. All teaching should be for understanding and not for parrot learning.

9.A No. Ears are important, but our eyes give us most of our experience, particularly in technical training.

B *Yes.* About 75 per cent of our experience comes to us through our eyes. 'A picture is worth a thousand words.'

C No. Encourage learning by doing, not feeling with the hands.

D No. Does not play a major part in the acquisition of learning by experience.

10.A No. Otherwise a taped recording would be equally effective.

B *Yes.* Use relevant audio-visual aids as often and effectively as possible.

C No. The maximum class participation should be aimed at in all technical training. Use senses of sight, touch and hearing to maximum advantage.

D No. The appeal to the eye should reinforce the words to facilitate understanding and assimilation.

11.A No. A diagram is largely meaningless unless the real thing has been seen.

B No. Sight of the equipment should not be a treat at the end of a boring talk, it is an essential introduction.

C *Yes.* Introduce the whole thing first; use senses of sight, hearing and feeling. Satisfy curiosity and stimulate interest.

D No. Could be dangerous for students and machines, and wrong notions are difficult to eradicate.

12.A No. They probably know the answers and are not in special need of stimulation.

B No. They will feel they are being got at, or shown up before the others, and may be discouraged.

C No. May increase their nervousness if they are not treated like the rest of the class.

D *Yes.* This gives the instructor a good idea of class progress and discloses weaknesses in individuals and in instruction. It is fair, and no one can anticipate being asked, or not being asked, a question.

13.A No. Questions may help of course but they should be part of instruction, not merely anti-sleep stimulant. Find out why the class want to sleep! Poor preparation? Presentation? General instruction?

B No. Could have a limited use for this purpose, but questions should encourage full class participation.

C *Yes.* Essential in all good instruction; ensures contact with the class and stimulates thought and interest.

D No. This is a subsidiary function. *All* students need encouragement.

14.A *Yes.* This helps class participation, and ensures a sound teacher–class relationship.

B No. Questions to and from students are an essential part of instructions at all stages of the lesson.

C No. Detracts from a feeling of active individual participation in the lesson. Questions and answers should be heard by all, unless you are running a brains trust which requires written questions.

D No. Questions should be encouraged. They indicate how the lesson is going, and what the student's difficulties are.

15.A No. Does not encourage the middle and lower sections of the class to think out the answers.

B No. Does not stimulate the rest of the class. All except the student addressed will tend to feel a sense of relief at not being asked.

C *Yes.* Ensures that each member of the class attends to the question and thinks of an answer in anticipation of being asked.

D No. Does not really save time. Unless the class is thinking and participating, the lesson is ineffective and the whole period is wasted. Wrong answers are useful in themselves as an indication of individual difficulties.

16.A No. This deceives no one.
B No. Indicates that you do not know, and want time to look it up.
C *Yes.* This is honest, and the class will respect you, but be sure to find out the answer as soon as possible.
D No. This is bluff; the class will detect it as such and your standing will be lowered.

17.A No. The question may have been very easy, or the marking very liberal.
B *Yes.* A good examination should show a spread of marks. Here they are compressed between 70 per cent and 95 per cent.
C No. A possibility, but not a reasonable conclusion. There is a poor spread in the marks as all seem to have done very well. Evidently something was wrong with the examination.
D No. There is no evidence that this has happened.

18.A No. Could be a proper conclusion only if the test had been standardised or the same objective test had been taken by each class.
B No. The marking may have been more strict.
C No. The examination may have been harder or the marking less liberal.
D *Yes.* No conclusion is possible from these results, without a great deal more information. Mere marks should not be endowed with properties they do not, and were not intended to, possess.

19.A Not necessarily. Each examination may have been easier or marked more leniently. The trainee might have been top of the class with 46 per cent or bottom of the class with 55 per cent.
B No. Cannot be inferred from the marks.
C *Yes.* No conclusion can be drawn without much more information about the examination, or the trainees position in the class, after each class test.
D No. If obtained in a hard test, and strictly marked they could represent a high standard of achievement.

20.A No. By the evening mental and physical fatigue begins to be felt.
B *Yes.* Generally the best time for a difficult piece of work.

C No. By the end of the morning students are more tired than at the beginning.

D No. Not the best time of day for mental exertion.

21.A *Yes.* A man works best when he knows how he is getting on. He is motivated by regular progress reports.

B No. Can have the effect of discouraging the class. Praise is better than blame as a motivator.

C No. Better than 'B' but may lead to complacency.

D No. This method tends to discourage a class and will not get the best out of them.

22.A No. Stable groups are valuable in training.

B *Yes.* Provided the stable groups are reasonably sound, with no really undesirable characters in them.

C No. The best types are valuable in making stability.

D No. Will tend to spread instability unnecessarily.

23.A No. A trainee wants to know where he stands as soon as he can be told.

B *Yes.* Fair, honest and can be helpful to the trainee. At the end of the interview give him every possible encouragement and guidance.

C No. General exhortations are of little use. Be specific in letting a man know what he must do to improve.

D No. Spreads false optimism and may not be honest.

24.A *Yes.* Helps to create confidence and respect, and helps to build good personal relationships.

B No. The line of least resistance. Things may not right themselves and the trainee will feel you are not interested in him.

C No. May discourage a man and could do a great deal of harm. As you have not tried to find out what is wrong, you will be acting very unfairly.

D No. Find out what is wrong before taking any kind of action. This particular course would work only if done with sympathy and understanding.

25.A No. Useful but not the *first* thing to do.

B No. But it is perhaps the second thing to have on your list.

C *Yes.* You are liable to waste time and effort unless you know exactly what you are expected to achieve.

D No. Will be useful to you but what questions would you ask them? You must have some clear ideas yourself.

26.A No. Will be boring if the class already knows a good deal about the subject.
B No. Depends on your making a guess, and you may be wrong.
C *Yes.* Gives you valuable initial information about the knowledge of the class, and about individuals, that will be useful to you in preparing the course.
D No. Might waste several weeks. You must know what stage your class is at before you start.

27.A No. A film should be shown as a part of a course, at the appropriate place.
B No. Fails to recognise a film as an important training aid. It encourages the wrong but popular belief that a film is a form of entertainment and a pleasant relaxation requiring no effort.
C No. A film is not merely a stimulant, or something to restore flagging interest.
D *Yes.* It is an audio-visual aid to be used as a part of a course. Discuss it before and after showing.

28.A No. Most unlikely that the others will derive any real benefit from this. The matter is too important to be left to chance.
B No. Has the disadvantage that in each class the best students will be working at a slower pace than they could. It is wrong to have the best section of your class working at less than full capacity.
C *Yes.* Will generally be found to be the best solution. Each class will go at its own maximum pace. Students of outstanding ability in a mixed group may be a source of discouragement to the very poor students rather than a stimulus.
D No. Details are important and should not be left to chance.

29.A No. Demonstration is important and valuable but should not occupy most of the time. How could you teach a boy to ride a bicycle?
B No. Too small a proportion of the time for practical work. How did you learn to ride a bicycle?
C No. Practical work in training for a skill should occupy more than half the time allotted.

D *Yes.* A skill can be taught effectively by allocating about 10 per cent of the time to explanation, 25 per cent to demonstration, and 65 per cent of the time to practice.

30.A No. Not the main reason. Teaching machines are not intended to replace the instructor.

B No. It should not affect the allocation of time to be spent on practical work.

C *Yes.* In addition, getting the right response is a reward and motivates the trainee.

D No. The machine should encourage all students by the graded steps in the programme, and not only the best ones.

Appendix II

If you have found the questions in Appendix I rather simple and requiring little thought, here is a set of a somewhat more exacting level, designed to provide examples of the use of this type of question, not so much as a means of testing a person's knowledge, but more as a means of stimulating thought and providing a basis for a discussion of some of the principles and points of view dealt with in the book. The questions have been selected with some adaptations from the book *Test Items in Education* (3 100 test items) by C. L. Mouly and L. E. Walton through the courtesy of the publishers Shaum Publishing Co., New York.

In many of the questions the 'book' answer given is open to question as a number of the questions have more than one good or acceptable answer among the choices offered. Where this occurs the person trying the test is invited to select the answer which seems to be the best, or the most helpful in the learning process. Discussions on some of the alternatives offered will give ample support to the idea that learning comes from participation even though this means defending one's position. Testing and feedback are part of the learning process.

1. Upon which of the following does the skill of roller-skating most clearly depend?

(*a*) development of inherited potential;
(*b*) development of acquired potential;
(*c*) training almost exclusively;
(*d*) training superimposed upon readiness to learn;
(*e*) interest and opportunity for training.

2. Which of the following statements is false? Research on the correlation among abilities has indicated that:

(*a*) a person weak in one area will generally have at least one area in which he can excel;

(*b*) a person weak in one area will generally have compensatory superiority in other areas;
(*c*) an adult with all-round ability is the exception rather than the rule;
(*d*) the inter-correlation among traits measured by common tests is generally positive but low;
(*e*) the correlation between mental and physical traits decreases from infancy to adulthood.

3. The best method of dealing with the adolescent who is very awkward is:

(*a*) to make good-natured fun of his awkwardness;
(*b*) to be mildly but not overly sympathetic;
(*c*) to ignore him and his awkwardness until he attains greater co-ordination;
(*d*) to pressure him into practising the acts in which he appears awkward;
(*e*) to refer him for medical help.

4. Good classroom discipline is best recognised by:

(*a*) the order and quiet prevailing in the classroom;
(*b*) the learning going on in the classroom;
(*c*) the co-operation and group feeling among members;
(*d*) the teacher–pupil rapport that prevails in and out of the classroom;
(*e*) the pupil growth it promotes.

5. Which of the following would you consider most basic as a cause of teacher failure in discipline?

(*a*) lack of understanding of adolescents;
(*b*) lack of understanding of the purpose of discipline;
(*c*) poor personality on the part of the teacher;
(*d*) lack of parental co-operation in the discipline of their children;
(*e*) lack of administrative support of the teacher's disciplinary measures.

6. Among the 'tips' for beginning teachers we should include:

(*a*) make discipline an end in itself;
(*b*) ignore petty violations that do not interfere with classwork;

(*c*) make sure children understand what you will do if they misbehave;
(*d*) prevent all misbehaviour so that good behaviour will become habitual;
(*e*) place all enforcement of rules in the hands of the pupils.

7. The punishment for a given offence should :

(*a*) be as severe as is necessary to prevent its recurrence;
(*b*) be set in advance for each type of offence;
(*c*) be related directly to the situation from which the trainee is being punished;
(*d*) be decided by the class and carried out without modification or relentment;
(*e*) have as its primary purpose the development of the individual.

8. The most valid indicator of good classroom discipline is :

(*a*) the absence of noise and disorder;
(*b*) the degree to which pupils master course content and skills;
(*c*) pupil conformity to rules and directives;
(*d*) the extent to which pupils accept responsibility for self-direction;
(*e*) pupil–teacher relationships.

9. The first step in the guidance of the learning of a complex skill is :

(*a*) to make sure the learner has a clear understanding of the goal to be attained;
(*b*) to provide the learner with a good demonstration of what is expected;
(*c*) to give the learner confidence in his ability to do what is expected;
(*d*) to warn the learner against the possible mistakes he might make;
(*e*) to have the learner practice the subskills before attempting the total performance.

10. In guiding the learning of a skill, it is generally best :

(*a*) to use verbal directions in the early stages and later to use demonstrations in the removal of errors;
(*b*) to let the learner learn as much as possible on his own (answer his questions but keep directions to a minimum),

(*c*) to concentrate on the removal of errors as they occur (do not anticipate their occurrence);
(*d*) to prevent the learner from making mistakes;
(*e*) to place the emphasis on 'what to do' rather than on 'what to avoid'.

11. In order to give effective guidance to the learning of a skill, the instructor must of necessity be able:

(*a*) to execute the skill with a high degree of perfection;
(*b*) to execute the skill with a higher degree of perfection than the person he is trying to instruct;
(*c*) to develop in the student insight into the nature of the skill to be learned;
(*d*) to foresee and forestall the development of bad habits, the occurrence of plateaus, etc.;
(*e*) to give slow-motion demonstrations of the skill without distortion.

12. What comes first in planning a course?

(*a*) determination of methods of evaluation and testing;
(*b*) determination of methods of presentation;
(*c*) determination of objectives;
(*d*) selection of the learning activities;
(*e*) selection of the materials of instruction.

13. The form of guidance most appropriate in the development of meanings, concepts, and other complex ideas is:

(*a*) manual;
(*b*) kinaesthetic;
(*c*) visual;
(*d*) verbal;
(*e*) a combination of the above in equal proportion.

14. Under most conditions, learning is best when:

(*a*) acquired in large amounts in a few long learning periods;
(*b*) broken up into the smallest possible units and the shortest possible practice periods;
(*c*) acquired by practice over distributed periods;
(*d*) adapted as to length and distribution of the learning periods to the nature of the specific situation;

(*e*) acquired in any amount which does not produce physical fatigue.

15. The principle of practice or exercise, as presently interpreted, postulates that:

(*a*) it is the proper kind of practice—rather than practice *per se*—that leads to learning;
(*b*) practice, to be effective, must be overt;
(*c*) the thoroughness of learning is proportional to the amount of practice put into it;
(*d*) practice is more effective in the removal of errors than in consolidation of the correct response;
(*e*) practice is effective in promoting learning only in instances which have a physiological basis.

16. Which of the following is generally the best organisation of practice periods in order to promote effective learning?

(*a*) long practice periods separated by long intervals between practice;
(*b*) long practice periods separated by short intervals between practice;
(*c*) short practice periods separated by moderately short intervals between practice;
(*d*) short practice periods with progressively longer intervals between practice.

17. The most effective type of practice is that in which the learner:

(*a*) is completely on his own;
(*b*) uses his learnings in the solution of more advanced problems;
(*c*) attains a high degree of perfection;
(*d*) is carefully prevented from making mistakes;
(*e*) is relatively unconcerned as to the outcome.

18. Which of the following is probably the most convincing explanation of the superiority of short and spaced practice periods over cramming?

(*a*) it allows different approaches on successive practice periods;
(*b*) it avoids fatigue;
(*c*) it minimises loss involved in warm-up periods;

(*d*) it allows wrong responses to drop out;
(*e*) it permits the right response to consolidate.

19. Research in the relative effectiveness of learning by wholes and by parts has shown that:

(*a*) learning by parts is better than by wholes;
(*b*) learning by wholes is better than by parts;
(*c*) the relative effectiveness of each depends on the nature of the material being learned;
(*d*) the relative effectiveness of each depends on the nature of the learner and of the material;
(*e*) learning by wholes tends to be more effective in most cases.

20. Probably the most effective way of learning a complex and complicated skill is:

(*a*) to practice the whole skill over and over;
(*b*) to practice each part of the skill separately;
(*c*) to practice the whole skill with separate attention to parts where indicated;
(*d*) to alternate between practice on the whole and on the parts;
(*e*) to integrate part by part into a continuous performance before proceeding to the next part.

21. The major argument against learning by wholes is that:

(*a*) it may lead to an underlearning of some of the parts and over-learning of others;
(*b*) it does not give sufficient reinforcement to right and wrong responses;
(*c*) it is not as efficient as the part method;
(*d*) it creates anxiety in young and insecure learners;
(*e*) its superiority to the part method is restricted to meaningless material.

22. Purposeful group learning in the classroom generally begins with:

(*a*) intelligent trial-and-error activity;
(*b*) motives and a clarification of objectives;
(*c*) hypotheses and generalisations;
(*d*) delegation of responsibility to the members of the group;
(*e*) exposition of the plan of attack by the teacher.

23. The rules of mensuration in arithmetic are most effectively learned through :

(*a*) systematic drill;
(*b*) incidental everyday experiences;
(*c*) specially selected meaningful experiences;
(*d*) a topic-by-topic approach to the study of arithmetic;
(*e*) a systematic procedure of laying out and measuring each geometric figure.

24. A teacher in introducing a new subject upon meeting his class for the first time had best :

(*a*) begin with the first lesson without delay;
(*b*) give the class a broad outline of the subject;
(*c*) begin at once with the review of the relevant material of the previous grades;
(*d*) concentrate on identifying potential troublemakers and leaders of classroom mischief;
(*e*) divide his class into teams for discussion, group projects, etc.

25. Notes from lectures or readings should include :

(*a*) everything time permits;
(*b*) only the major generalisations;
(*c*) the main points with enough supportive data to preserve the meaning;
(*d*) very little, the time is better spent in concentrating on the lecture or readings;
(*e*) nothing but a few key words to mark the general direction of the discussion.

26. Research evidence indicates the most effective placement of study questions is generally :

(*a*) at the end of the learning;
(*b*) at the beginning of the lesson;
(*c*) during the last half of the lesson;
(*d*) a matter of personal opinion and preference; the research is inconclusive;
(*e*) determined by the background already possessed by the learner.

27. Which of the following would be characteristic of good notes?

(*a*) they are taken as you go along reading the material;
(*b*) they are taken in the words of the author;
(*c*) they are thorough and comprehensive with a maximum of detail;
(*d*) they are in outline form with a minimum of supportive detail;
(*e*) they are arranged in a bound notebook with entries arranged in chronological order.

28. The most important element in instruction is the:

(*a*) variety of instructional materials used;
(*b*) teaching techniques used;
(*c*) subject-matter covered;
(*d*) relationship between teacher and student;
(*e*) evaluation of learning.

29. The drilling of facts:

(*a*) has no place in the modern training curriculum?
(*b*) is valuable if used for a certain purpose;
(*c*) should be used extensively in every classroom;
(*d*) should be employed only as a last resort;
(*e*) provides an excellent rest period for the teacher.

30. Concerning the lecture technique, the consensus is that the lecture:

(*a*) has no place in teaching today;
(*b*) is a good way to ensure learning;
(*c*) has a motivational function;
(*d*) functions best when used alone;
(*e*) should be used exclusively.

31. The lecture method would be most appropriate when used to:

(*a*) hold the attention of an unruly class;
(*b*) develop interest and supplement the text;
(*c*) ensure complete coverage of material;
(*d*) teach factual subjects;
(*e*) teach controversial topics.

32. Which of the following are classified under audio-visual or visual aids?

(*a*) sound motion pictures and television;

(*b*) pictures, drawings and diagrams;
(*c*) radio, phonograph records and slides;
(*d*) tape recordings, maps and cartoons;
(*e*) all of the above are correct.

33. Audio-visual aids should be considered by the teachers as a :

(*a*) substitute for traditional teaching;
(*b*) pleasant change from classroom routine;
(*c*) supplement to other devices for teaching;
(*d*) device which saves teacher preparation;
(*e*) solution to the problem of teacher shortage.

34. Pupil participation is considered superior to lecture-recitation procedures mainly because of the :

(*a*) number of facts students learn;
(*b*) amount of work students do;
(*c*) quality of the learning derived from such activity;
(*d*) amount of teacher ingenuity and preparation required;
(*e*) superior scores made on tests.

35. Which of the following is most basic in determining the effectiveness of a given lesson?

(*a*) the appropriateness of the audio-visual aids used;
(*b*) the adequacy of the teacher's presentation of the topic;
(*c*) the classroom atmosphere and the behaviour of the pupils;
(*d*) the logical organisation of the subject matter;
(*e*) the degree of pupil involvement in the learning activity.

36. The major goal in classroom discipline should be to promote :

(*a*) quiet and order in classroom and study hall;
(*b*) good student conduct;
(*c*) the complete elimination of pupil misconduct;
(*d*) self-direction on the part of the pupils;
(*e*) good study conditions.

37. The chief cause for failure of most teachers in discipline is :

(*a*) soft-heartedness;
(*b*) failure to clarify the rules they enforce;
(*c*) inability to understand students;
(*d*) lack of knowledge of the subject taught;
(*e*) inconsistent action.

A Test on Tests

38. The attitude concerning tests which is most consistent with modern educational theory is:

(*a*) a standardised test is the only valid and reliable test for a teacher to use;
(*b*) a test is but one of several devices for appraising the progress of pupils;
(*c*) only objective measures should be used in appraising pupil progress;
(*d*) new type tests are better than essay examinations for all appraisal purposes;
(*e*) only subjective measures should be used in appraising pupil progress.

39. The type of test or test item to be used should be decided mainly on the basis of the:

(*a*) necessity for variety in the tests;
(*b*) nature of the skill or knowledge to be tested;
(*c*) availability of standardised tests;
(*d*) ease with which it can be constructed;
(*e*) length of time spent on the topic.

40. A recall situation is best tested by the:

(*a*) multiple-choice type test;
(*b*) completion type test;
(*c*) true–false type test;
(*d*) topical-selection type test;
(*e*) multiple-response type test.

41. Within limits, a standardised test has, as its chief virtue, the fact that it:

(*a*) is easier to administer than a teacher-made test;
(*b*) is generally of a higher quality than a teacher-made test;
(*c*) permits a comparison of individual and class performance with other classes and individuals;
(*d*) has more ceiling, thereby permitting the brighter students to display what they have learned;
(*e*) is easier to correct and grade than a teacher-made test.

42. The term 'mean' refers to :

(*a*) the most frequently made score;
(*b*) the middle score in a distribution;
(*c*) the arithmetic average of all scores;
(*d*) the score representing the passing grade;
(*e*) the highest score made on the examination.

43. Which of the following is most acceptable evidence of the validity of the test?

(*a*) the test covers a wide variety of topics;
(*b*) average students tend to obtain average scores on the test;
(*c*) the test scores form a bell-shaped curve;
(*d*) the items bears directly upon the objectives of instruction.

44. In checking the discriminatory power of test items, the teacher can find out :

(*a*) the validity of the test only;
(*b*) the usability of the test only;
(*c*) the objectivity and reliability of the test;
(*d*) the difficulty level and validity of the test;
(*e*) none of the above.

45. Which of the following statements concerning test validity and reliability is most accurate?

(*a*) a test cannot be valid unless it is reliable;
(*b*) a test cannot be valid unless it is objective;
(*c*) a test cannot be reliable unless it is valid;
(*d*) a test cannot be reliable unless it is standardised;
(*e*) a test cannot be standardised unless it is reliable.

46. Which of the following best describes the factor of validity in a test?

(*a*) accurate measurement of the ability to be appraised;
(*b*) consistent measurement of whatever ability the test measures;
(*c*) freedom from subjective (chance) factors;
(*d*) internal consistency of the items of the test;
(*e*) standardisation of administrative routine.

47. Other things being equal, which type of test tends to have the lowest reliability?

(*a*) completion;

(*b*) essay;
(*c*) matching;
(*d*) simple recall;
(*e*) true–false.

48. The ideal difficulty of the items of a teacher-made test is such that:

(*a*) every item is passed by approximately 50 per cent of the class;
(*b*) the average student gets 50 per cent of the maximum score possible;
(*c*) the average score for the class is in the neighbourhood of 75 per cent of the maximum score possible;
(*d*) few, if any, students get less than 50 per cent of the maximum score possible;
(*e*) a minimum of one or two students get near-perfect scores on the test.

49. As a rough rule, the time limits of a test should be set so that approximately .... per cent of the students finish the test in the time allowed.

(*a*) fifty;
(*b*) sixty-five;
(*c*) eighty;
(*d*) ninety;
(*e*) ninety-eight.

50. A student obtained a score of 72 on a test. This means that:

(*a*) he has exceeded in performance 72 per cent of his fellow-students;
(*b*) he answered correctly 72½ of the items on the test;
(*c*) he answered 72 items on the test;
(*d*) he ranked 72nd in the class;
(*e*) he obtained a raw score of 72—and little else.

51. The percentile rank of a given score deals with the per cent of the cases of a distribution .... the given score.

(*a*) at or above;
(*b*) above;
(*c*) at;
(*d*) below;
(*e*) at or below.

52. Q1 is to the 25th percentile as the median is to:

(a) the 40th percentile;
(b) the 50th percentile;
(c) the 60th percentile;
(d) the 75th percentile;
(e) none of the above.

53. Because a test was too hard for a class, each student's score is increased by a factor of 1·25. Which of the following would not be affected by the bonus?

(a) the class mean;
(b) the class median;
(c) the range of scores in the class;
(d) the class standard deviation;
(e) the correlation of the grades and IQ's of the students of the class.

54. The measure of central tendency to avoid if, for any reason, one wishes to avoid the influence of extreme scores is:

(a) the mean;
(b) the median;
(c) the mode;
(d) the quartile;
(e) the standard deviation.

55. The greatest weakness of the range as a measure of variability is:

(a) its complicated computation;
(b) its difficulty of interpretation;
(c) its extreme instability;
(d) its violation of the rules of mathematics;
(e) its lack of independence of the distribution of the variable involved.

56. The variance is best thought of as:

(a) a term essentially synonymous with variate or variable;
(b) an important source of variability;
(c) the square of the standard deviation;
(d) the average variability of the scores of a given distribution;
(e) a technique for appraising the homogeneity of the scores of a distribution.

57. The standard deviation is best thought as :

(*a*) a measure of the degree of relationship between two variables;
(*b*) a standard or average error of the items in tabulated data;
(*c*) the midpoint of the 16th and the 84th percentile;
(*d*) the square root of the mean of the squares of the deviations about the mean;
(*e*) a measure used to specify half the spread of the middle 50 per cent of the cases of a given distribution.

58. Statisticians consider which of the following as the best all-round measure of variability?

(*a*) the range;
(*b*) the average deviation;
(*c*) the quartile deviation;
(*d*) the standard deviation;
(*e*) the P_2 to P_{98} range.

59. When a distribution of achievement test scores is positively skewed, one might infer that the test was :

(*a*) too easy;
(*b*) too hard;
(*c*) unreliable;
(*d*) poorly standardised;
(*e*) incorrectly administered.

Answers for consideration

1. *d*.	14. *d*.	27. *d*.
2. *b*.	15. *a*.	28. *d*.
3. *b*.	16. *c*.	29. *b*.
4. *e*.	17. *b*.	30. *c*.
5. *a*.	18. *d*.	31. *b*.
6. *b*.	19. *c*.	32. *e*.
7. *c*.	20. *c*.	33. *c*.
8. *d*.	21. *a*.	34. *c*.
9. *a*.	22. *b*.	35. *e*.
10. *a*.	23. *c*.	36. *d*.
11. *c*.	24. *b*.	37. *c*.
12. *c*.	25. *c*.	38. *b*.
13. *d*.	26. *b*.	39. *b*.

40. *b.*
41. *c.*
42. *c.*
43. *d.*
44. *d.*
45. *a.*
46. *a.*
47. *b.*
48. *b.*
49. *c.*
50. *e.*
51. *e.*
52. *b.*
53. *e.*
54. *a.*
55. *c.*
56. *c.*
57. *d.*
58. *d.*
59. *b.*

Bibliography

ANNETT, J. (1969). *Feedback and Human Behaviour.* Penguin, Harmondsworth, Middlesex.

ATKINSON, N. J. (1966). *Modern Teaching Aids.* Maclaren and Sons, London.

BAGLEY, W. A. (1962). *Facts and How to Find Them.* Pitman, London.

BARBER, J. W. (Ed.) (1968). *Industrial Training Handbook.* Ilife.

B.B.C. (1971). *Leaflet on Video Tape Recording.*

BEARD, R. (1970). *Teaching and Learning in Higher Education.* Penguin, Harmondsworth, Middlesex.

BELBIN, EUNICE. (1964). *Training the Adult Worker.* H.M.S.O., London.

BELBIN, R. M. (1969). *The Discovery Method in Training.* H.M.S.O., London.

BEVERIDGE, A. (1963). *Apprenticeship Now: Notes on the Training of Young Entrants to Industry.* Chapman and Hall, London.

BEVERSTOCK, A. G. (1968). *Industrial Training Practice.* Classic Publications.

BILODEAU, E. A. (1969). *Principles of Skill Acquisition.* Academic Press, New York.

BISHOP, M. H. (1969). *Dynamic Supervision.* American Management Association.

BLACK, J. M. (1970). *How to Get Results from Interviewing: A Practical Guide for Operating Management.* McGraw-Hill, New York.

BLUM, M. L., and NAYLOR, J. C. (1970). *Industrial Psychology.* Harper and Row, London.

Books for Training Officers (1969). BACIE and National Book League, London.

BORGER, R. and SEABOURNE, A. B. M. (1970). *The Psychology of Learning.* Penguin, Harmondsworth, Middlesex.

BOYDELL, T. H. (1970). *A Guide to Job Analysis.* BACIE, London.

British Association for Commercial and Industrial Education. *Training Aids Index.* BACIE, 16, Park Crescent, London.

BRITISH INSTITUTE OF MANAGEMENT. *A Basic Library of Management.* Management Publications Ltd., London.

BROADWELL, M. M. (1971). *The Supervisor as an Instructor*, 2nd edn. Addison Wesley, Reading, Mass.

BROWN, J. (1966). *Objective Tests—Their Construction and Analysis.* Longmans, London.

BRUTON, E. (1962). *Automation.* Muller, London.

CABLE, R. (1968). *Audio Visual Handbook.* University of London Press, Ltd., London.

COMPTON, H. P. (1962). *Conveying Ideas: Fundamentals of Written Communication.* Macmillan, London.

CROCKER, A. C. (1969). *Statistics for the Teacher.* Penguin, Harmondsworth, Middlesex.

DAVIES, I. K. (1971). *The Management of Learning.* McGraw-Hill, New York.

DENERLEY, R. A. and PLUMBLEY, P. R. (1968). *Recruitment and Selection.* Institute of Personnel.

DENT, H. C. (Ed.) (1971). *The Year Book of Technical Education and Training for Industry.* A. and C. Black, London.

DEPARTMENT OF EDUCATION AND SCIENCE. (1965). *Audio-Visual Aids in Higher Scientific Education.* H.M.S.O., London.

DEPARTMENT OF EMPLOYMENT (1970). *Training Research Register 1970–71.* H.M.S.O., London.

DEPARTMENT OF EMPLOYMENT. (1970). *Basic Rules for Safety and Health at Work.* (*Health and Safety at Work Series, Booklet* No. 35). H.M.S.O., London.

DEPARTMENT OF EMPLOYMENT AND PRODUCTIVITY (1970). Government role in industrial training. *Employment and Productivity Gazette*, April.

DEPARTMENT OF EMPLOYMENT (1971). *Glossary of Training Terms.* H.M.S.O., London.

D.S.I.R. (1960). *Training made Easier.* H.M.S.O., London.

EDUCATIONAL FOUNDATION FOR VISUAL AIDS. (19??). *Choosing a Tape Recorder.* National Committee for Audio-Visual Aids in Education, 38 Queen Anne St., London.

D.S.I.R. *The Use of Audio-Visual Aids in Colleges of Further Education.* Educational Foundation for Visual Aids, London.

ELLIOT, F. (1964). *A Technical Teachers' Certificate Course.* Cassell, London.

ERANT, MICHAEL, and SQUIRES, GEOFFREY. (1970). *An Annotated Select Bibliography of Educational Technology.* National Council for Educational Technology. Books for Schools Ltd., London.

GRANT, M., and SQUIRES, G. (1971). *An annotated Select Bibliography of Educational Technology.* National Council for Educational Technology.

FELDMAN, M. P. (1971). *Psychology in the Industrial Environment.* Butterworths, London.

FIELDS, A. (1969). *Method Study.* Cassell Management Studies, London.

FINNIGAN, J. (1970). *Industrial Training Management.* Finnigan J. Business Books Ltd.

FRAZER, J. MONROE. (1966). *A Handbook of Employment Interviewing.* Macdonald and Evans.

FRASER, J. MUNRO, and BRIDGES, J. M. (1964). *The Industrial Supervisor.* Business Publications, London. (1963).

GEB BLOOM, B. S. (1956). *Taxonomy of Educational Objectives.* Longmans Green, Harlow, Essex.

GOLDSMITH, M. (1963). *Mechanisation in the Classroom.* Souvenir Press, London.

GOUGH, J. S. (1970). *Interviewing in Twenty-Six Steps.* BACIE, London.

HANDLEY, W. (Ed.) (1969). *Industrial Safety Handbook.* McGraw-Hill, London.

HANDY, H. W., and HUSSAIN, K. M. (1969). *Network Analysis for Educational Management.* Prentice-Hall, London.

HARE, R. (1970). *Know How—A Student Guide to Project Work.* International Textbook Company, Insight Series, London.

HEYEL, C. (4th ed.) (1971). *The Foreman's Handbook.* McGraw-Hill, New York.

HUGHES, A. G. and E. H. (1959). *Learning and Teaching.* Longmans, London.

HUGHES, E. W. (1970). *Human Relations in Management.* Pergamon Press, Oxford.

JONES, C. (1970). The role of films in training. *Industrial Training International,* May 5.

KAY, H., ANNETT, J., and SIME, M. E. (1963). *Teaching Machines and their Use in Industry*. H.M.S.O., London.

KING, D. (1964). *Training within the Organization: A Study of Company Policy and Procedures for the Systematic Training of Operators and Supervisors*. Tavistock.

KITSON CLARK, G., and BIBBER CLARKE, E. (1959). *The Art of Lecturing*. Heffer, Cambridge.

KROLL, A. M., DINKLAGE, L. B., LEE, J., MORLEY, E. D., and WILSON, E. H. (1970). *Career Development: Growth and Crises*. Wiley, New York.

LAMB, B. (1970). *New Methods and Media in Further Education*. National Committee for Audio-Visual Aids in Education, London.

LAMB, R. T. B. (1967). *Aids to Modern Teaching—A Short Survey*. Pitman, London.

LIEPMAN, KATE. (1960). *Apprenticeship: An Inquiry into its adequacy under modern conditions*. Routledge, London. (1964).

LOVELL, K. (1967). *Team Teaching*. Univ. of Leeds Institute of Education, Leeds.

MACLENNAN, A. (1964). *Technical Teaching and Instruction*. Oldbourne, London.

MARTIN, J. and NORMAN, A. R. D. (1970). *The Computerized Society*. Prentice-Hall, London.

MARKLE, S. (1969). *Good Frames and Bad*. Wiley, New York.

MEREDITH, G. P. (1961). *Learning, Remembering and Knowing*. English Universities Press, London.

METRICATION BOARD. (1970). *Going Metric—The First Five Years 1965–69*. H.M.S.O., London.

METRICATION BOARD (1971). *Going Metric—Progress in 1970*. H.M.S.O., London.

SI in Engineering, Council of Technical Examining Bodies, Sales Section, City & Guilds of London Institute, 76 Portland Place, London WIN 4AA.

MILWARD, G. E. (Ed.) (1967). *Organisation and Methods*. Macmillan, London.

MORRIS, F. C. (1950). *Effective Teaching: A Manual for Engineering Instructors*. McGraw-Hill, New York.

NATIONAL COUNCIL FOR EDUCATIONAL TECHNOLOGY. (1969). *Towards More Effective Learning*. (The Report of the National

Council for Educational Technology 1967–68). Education Press, London.

NICKSON, MARYLYN. (1971). *Educational Technology: A Systematic Approach for Teachers.* Ward Lock, London.

NORMAN MACKENZIE GRANT. *Teaching and Learning: An introduction to New Methods and Resources in Higher Education.* H.M.S.O., London, for Unesco IAU.

ODIORNE, G. S. (1970). *Training by Objectives—An Economic Approach to Management Training.* Collier-Macmillan, London.

PAUL, W. J., and ROBERTSON, K. B. (1970). *Job Enrichment and Employee Motivation.* Gower Press, London.

PINSENT, A. (1962). *Principles of Teaching Method.* Harrap, London.

PLATT, B. M. (Ed.) (1967). *Films for Managers: A Guide to Management and Supervisory Training Films In the UK.* Management Publications, British Institute of Management.

POWELL, L. S. (1969). *Communication and Learning.* Pitman, London.

POWELL, L. S. (1970). *A Guide to the Overhead Projector.* BACIE, London.

POWELL, L. S. (1970). *A Guide to the Use of Visual Aids,* 3rd edn. BACIE, London.

POWELL, L. S. (1970). *Lecturing to Large Groups,* 3rd impression. BACIE, London.

POWELL, G. H. and L. S. (1967). *A Guide to the 8 mm Loop Film.* BACIE, London.

PRIESTLEY (1971). *The Industrial Training Yearbook.* 1971. Kogan Page, London.

RICHMOND, W. K. (1967). *The Teaching Revolution.* Methuen, London.

RIGG, R. P. (1969). *Audio-visual Aids and Techniques in Managerial and Supervisory Training.* Hamish Hamilton, London.

ROBINSON, J., and BARNES, N. (Eds.) (1968). *New Media and Methods in Industrial Training.* B.B.C., London.

ROGERS, J. (Ed.) (1969). *Teaching in equal terms: a book for all those Concerned with Teaching Adults.* B.B.C., London.

ROGERS, J. (1970). *Using Broadcasts—A Guide for Tutors of Adult Groups.* B.B.C. Radio Television Publications, London.

SEYMOUR, W. D. (1954). *Industrial Training for Manual Operators.* Pitman, London.

SEYMOUR, W. D. (1966). *Industrial Skills.* Pitman, London.

SINGER, E. J., and RAMSDEN, J. (1969). *The Practical Approach to Skills Analysis.* McGraw-Hill, New York.

SKINNER, B. F. *The Technology of Teaching.* Appleton–Century–Crofts, New York.

SMITH, P. (1970). *Improving Skills in Working with People. The T. Group.* H.M.S.O., London.

TANSEY, P. (1970). Simulation—its role in training. *Industrial Training International*, Aug./Sept. 1970.

TAYLOR, E. A. (1966). *Visual Presentation in Education and Training.* Pergamon Press, Oxford.

THURLEY, J. E. and HAMBLIN, A. C. (1963). *The Supervisor and his Job.* H.M.S.O., London.

TURNER, P. and ATKINSON, C. R. M. *An Experiment in Closed-Circuit Television at Millfield School.* National Committee for Audio-Visual Aids in Education, London.

UNESCO. (1970). *Source Book for Science Teaching.* Unesco Educational Productions. UNESCO, Paris, VII.

UNGERSON, B. (Ed.) (1970). *Recruitment Handbook.* Gower Press, London.

VAN DERSAL, W. R. (1970). *The Successful Supervisor in Government and Business.* Pitman, London.

VINCENT, A. (1970). *The Overhead Projector (Experimental Development Unit Report No. 8).* National Committee for Audio-Visual Aids in Education.

WELLENS, J. (1963). *The Training Revolution: From Shop-floor to Board-room.* Evans.

Yearbook of Educational and Instructional Technology (1969–70). Cornmarket Press, London.

YOUNG, J. B. and BIRD, A. E. (1968). *Reprographic Principles Made Easy.* Educational Foundation for Visual Aids, London.

The Supervisor as an Instructor.

Careers Handbook.

Periodicals

Technical Education and Industrial Training (monthly). Evans Bros. Ltd., Montague House, Russell Square, London, W.C.1.

Industrial Training International (monthly). Newman Neame Ltd., 4 Fitzroy Square, London W.1.

BACIE Journal (quarterly). British Association for Commercial and Industrial Education, 16 Park Crescent, London W.1.

Visual Education (monthly). National Committee for Audio-Visual Aids in Education, 33 Queen Anne Street, London W.1.

Film User (monthly). Current Affairs Ltd., P.O. Box 109, Davis House 69–77 High Street, Croydon, Surrey.

Training for Progress (quarterly). Bulletin of the International Vocational Training Information and Research Centre. C.I.R.F., Geneva.

CETO News (quarterly). Centre for Educational Television Overseas, Nuffield Lodge, Regent's Park, London N.W.1.

Central Training Council publications (H.M.S.O.).

Journal of Educational Technology, National Council for Educational Technology.

Training Abstracts Service (monthly on cards). Information for Education Ltd., Walton Hall, Bletchley, Bucks.

British Journal of Educational Technology, National Council for Educational Technology Councils and Educational Press Ltd., 10 Queen Anne St., London, W.1.

Index